THE LEIGHTON BUZZARD LIGHT RAILWAY

and associated quarry lines

SYDNEY A. LELEUX

THE OAKWOOD PRESS
1969

C O N T E N T S

		Page
INTRODUCTION AND ACKNOWLEDGEMENTS.	.....	1
CHAPTER 1: LEIGHTON BUZZARD SAND.	.....	3
CHAPTER 2: THE LEIGHTON BUZZARD LIGHT RAILWAY.	.....	14
CHAPTER 3: LATER HISTORY.	.....	22
CHAPTER 4: THE DESCRIPTION OF THE ROUTE.	.....	26
CHAPTER 5: OPERATION.	.....	33
CHAPTER 6: LIGHT RAILWAY LOCOMOTIVES.	.....	40
CHAPTER 7: QUARRY SYSTEMS CONNECTED TO L.B.L.R.	.....	49
CHAPTER 8: QUARRY LOCOMOTIVES.	.....	65
CHAPTER 9: ROLLING STOCK.	.....	87
CHAPTER 10: IRON HORSE PRESERVATION SOCIETY.	.....	94
CHAPTER 11: QUARRY SYSTEMS NOT CONNECTED TO L.B.L.R.	.....	98

    1. Bedford Silica Sand Mines Ltd.
    2. H.G.Brown.
    3. Clay Cross Coal Depot.
    4. Grovebury Quarries.
    5. Leighton Buzzard Brick Co.Ltd.
       (a) Ledburn Road Pit.
       (b) Potsgrove Pit.
    6. Harry Sear.
    7. Woburn Abbey Parkland Railway.

---oooOooo---

INTRODUCTION AND ACKNOWLEDGEMENTS

INTRODUCTION AND ACKNOWLEDGEMENTS

In 1957 I visited Leighton Buzzard and made my first acquaintance with the local sand pit railways. Unfortunately I did not appreciate the historic interest of the locomotives then in use and made no notes. Later, it was suggested by a leading member of the Narrow Gauge Railway Society that I should write a history of the Leighton Buzzard Light Railway. This has taken about ten years intermittent work, during which time there have been great changes in the sand pit railways. Despite the slow progress of this history, I have found the research very rewarding.

Directors of the two major sand companies, J.E.Arnold of Joseph Arnold & Sons Ltd., and J.G.Delafield of George Garside (Sand) Ltd., have been very helpful, as has P.Cooke, Manager of Bedford Silica Sand Ltd. Manufacturers of railway equipment used, Robert Hudson (Raletrux) Ltd., Hudswell Clarke & Co.Ltd., and especially Motor Rail Ltd. have kindly given information as far as their records allow.

Documentary sources used are the "Leighton Buzzard Observer," the Memorandum and Articles of Association, together with a few balance sheets of the LBLR, Ministry of Transport returns, Ordnance Survey, especially the 1926 and 1937 editions of 25" maps (which account for the frequency of these dates in the text) the Victoria County History of Bedfordshire, "Light Railways of the First World War" by W.J.K.Davies, "The New Era Illustrated" of June 1930, and records at Shire Hall, Bedford.

Many railways and quarry employees have patiently answered my questions, amongst whom I would particularly like to mention A. Eggleton, C.Gaskin, G.Guess, S.Higgs, Mr.Jackson, T.Lambourne, H.G. and P.G.Lathwell, W.Major, C.Nash, F.G.North, F.C.Rickard, W."Old Bill" Smith, A.Tearle, and F.W.Turney. I am most grateful too, to those others whose names I do not know.

The publications and records of the Industrial Railway Society and Narrow Gauge Railway Society have been

INTRODUCTION AND ACKNOWLEDGEMENTS

of great use especially with regard to the locomotive stock. Members of these two Societies, P.R.Arnold, F.H.Eyles, F.Jux, A.Keef, R.Leleux, R.Peaman, P.Roberts, D.Semmens, G.H.Starmer, M.Swift, E.S.Tonks, B.Webb, and A.G.Wells have lent me their notes, observations and photographs.

At Leighton Buzzard I have been assisted by C.Daniels who has also done much fieldwork in response to my queries from Yorkshire. Without his help the final stages of preparation of this book would have been much more prolonged. He has been a very able and enthusiastic assistant.

My aunt Mrs. R. Higgs has always been generous in her hospitality on my visits to Leighton Buzzard.

Lastly I am greatly indebted to my wife who has first to endure piles of paper while I wrote, and then typed my untidy draft.

Keighley, Yorkshire.
July, 1969.

SYDNEY A. LELEUX

LEIGHTON BUZZARD SAND

Leighton Buzzard is a market town at the western end of Bedfordshire, 40 miles north of London. The market of Leighton was mentioned in the Survey of 1086. It has always been held on a Tuesday and stalls are still erected every week along the High Street. The ancient parish of Leighton Buzzard included the hamlets of Billington, Eggington, Heath, Reach and Stanbridge. These were made into civil parishes in the last quarter of the nineteenth century. The parish of Heath and Reach "is studded with sandpits, many of which are in use, the most extensive being in the south east corner at Shenley Hill. From one of these pits the sand used in the composition of the Crystal Palace was dug, owing doubtless to the influence of Sir Joseph Paxton, a native of this part of Bedfordshire and architect of the Crystal Palace." (Victoria County History Bedfordshire III 1912.)

The Lower Greensand (Lower Cretaceous period) which crosses Bedfordshire from south west to the north east near Potton has a maximum thickness of about 220' and an exposure several miles in width. "Near Leighton the Lower Greensand consists for the most part of white and light coloured sand known as 'silver sand' which has an industrial value. It is now chiefly obtained for filterbeds, but it is in places so free from iron and other colouring matter that it has been used for glassmaking. The larger quartz grains show signs of attrition, being rounded and polished, like the sands of the sea shore." (Victoria County History Bedfordshire I 1904). At Leighton a dark brown ferruginuous sandstone is also exposed. It had a limited use as building stone. This sandstone also occurs in isolated masses called "doggers". "Quite recently a fossilferrous band of sand was discovered at the summit of the series of pits which have been opened on Shenley Hill". The sand is found under fairly shallow cover, beneath a layer of blue clay of variable thickness and iron sandstone which often needs to be blasted.

LEIGHTON BUZZARD SAND

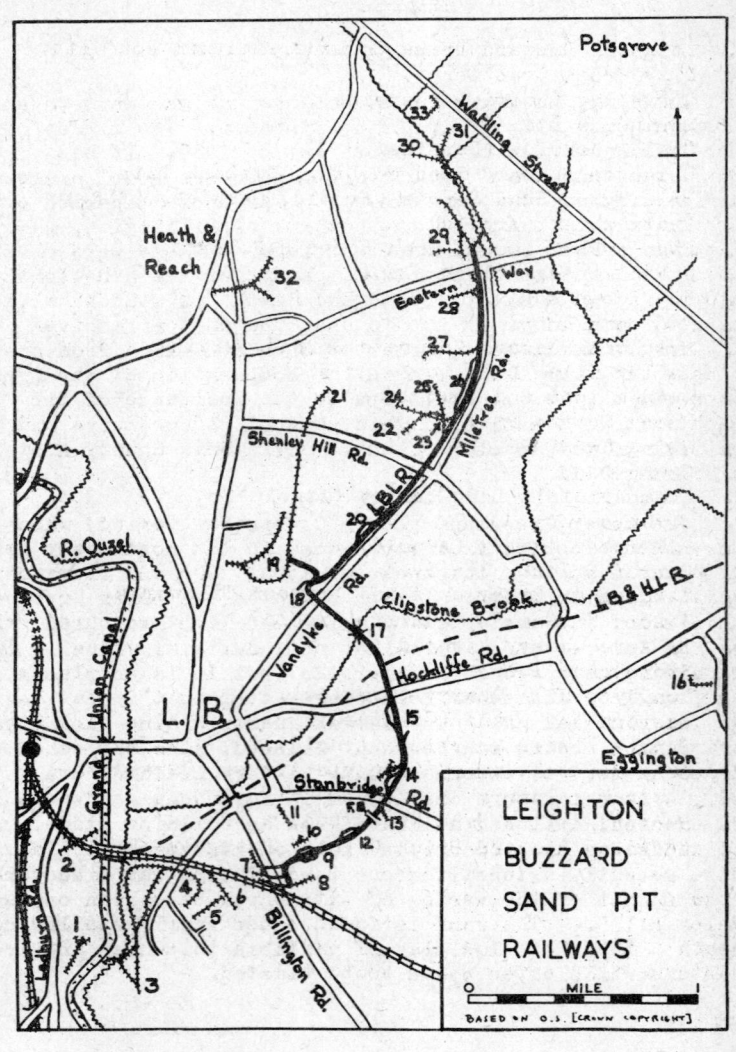

LEIGHTON BUZZARD SAND

MAP KEY

1. Leighton Buzzard Brick Co.Ltd., Ledburn Road Pit.
2. Clay Cross Coal Depot.
3. Grovebury Quarries (Garside).
4. Eastwoods Ltd.
5. Small quarries (Garside).
6. Leighton Buzzard Sand Co.Ltd., Firbank Pits.
7. Billington Road Depots (Arnold, Garside, LBLR.)
8. Pratt's Pit (Arnold).
9. Page's Park Loops, Iron Horse R.R.Depot.
10. H.Paul breeze block plant.
11. H.G.Brown's Pit.
12. Red Barn Loop.
13. Leighton Buzzard Concrete Co.Ltd. siding.
14. Marley Tiles Ltd. loop and siding.
15. Leedom loop and "dud loop".
16. Harry Sear's Pit.
17. Swing Swang Bridge.
18. Co-op Loop.
19. Chamberlain's Barn Quarry (Arnold).
20. Bryan's loop.
21. New Trees Quarry (Arnold).
22. Parrot & Jones Quarry.
23. Stonehenge Brickworks and Driroof Tileworks.
24. 9 Acre (Chance's) Quarry (Arnold).
25. 21 Acre Quarry (Arnold).
26. Stonehenge loop.
27. Munday's Hill Quarry (Garside).
28. Eastern Way plant (Garside).
29. Double Arches Quarries (Arnold).
30. Long Stretch Quarry - Double Arches (Garside).
31. Churchway Quarry   "   "   "
32. Bedford Silica Sand Mines Ltd.
33. Leighton Buzzard Brick Co.Ltd. Potsgrove Pit.

LEIGHTON BUZZARD SAND

A Silica and Moulding Sands Association publication describes the Lower Greensand deposits of Leighton Buzzard, Aylesbury, Redhill and King's Lynn as producing some of the best quality sands in the country. "The high silica content and grain characteristics of the deposits make them ideally suitable for core sands, resin-coated sands, and as the base sand for synthetic mixes for iron and steel moulding......the Greensand deposits of Southern England and East Anglia provide the glass and ceramic industries with almost the whole of the raw material which formerly (pre 1939) they had to import."

However, Leighton sand has not always been in a very competitive position. Before the First World War there was only a small profit in the sand trade due to foreign competition, especially from Belgian sand used as ship's ballast and then dumped in this country. The Leighton sand merchants sought to increase their profitability by reducing operating costs, especially transport costs. Even in 1930 foreign sand could be obtained in the north of England for less than the cost of rail carriage from Leighton Buzzard.

Sand had to be carried in carts from the pits, but the continuous heavy traffic damaged the roads and resulted in claims for compensation by the Bedfordshire County Council. Steam wagons were tried, as they offered some advantages, but were unkind to the macadamed surface. Around 1900 there had been plans for a railway to take the traffic off the roads, but nothing had materialised. Before 1914 the sand merchants' difficulties were acute.

The German invasion of Belgium in August 1914 cut off, almost overnight, supplies of Belgian foundry sand. At the same time the demand for such sand by the English munition factories was increasing rapidly. To alleviate the impending crisis, English sources of supply were sought and Leighton sand was found to be ideal. Production was increased and "enormous quantities" were sent daily by rail and canal to the munition centres.

LEIGHTON BUZZARD SAND

The carts and one or two steam tractors used to transport the sand from pit to the railway sidings at Grove Crossing were soon assisted by ten more tractors each hauling 5 to 10 ton loads of sand, 7 days a week. The roads began to take a fearful punishment and the Roads Board was compelled to take charge of the routes affected. Union Street (now Grovebury Road) cost as much in upkeep as Bedford High Street, and the monthly cost of road repairs due to the sand traffic rose to about £1000. Efforts were made to induce the Transport Board and Ministry of Reconstruction to build a light railway but to no avail. The cost of road repairs exceeded £25,000 in the war period.

With the end of this War the Roads Board handed back the sand routes to the county at the end of January 1919, and left the tractor owners to face the problems and liabilities of "extraordinary traffic". Their solution was to sell their equipment, the auction being held on Wednesday February 5th, 1919 at Messrs. Arnold & Sons' Union Street depot. The auctioneers were Messrs. Stafford, Rogers & A.W.Merry Ltd., and Cumberland & Hopkins. Heavy snow fell most of the time but "the 12 steamers, as they were lined up for sale, made an imposing show, each machine drawing up to the auctioneer's stand under full steam".

Haulage contractors and others from many parts of the midland counties attended the sale, and there was spirited bidding for many items. The ten 5 ton compound steam tractors, of various makes, with working pressures of 200 psi and upwards fetched about £550 each. Two traction engines were included in the sale, as well as wagons, spring trailers and accessories.

The sand carter came back into his own, but his swan song was brief, until the sand companies built their own light railway to reduce costs, which opened in November 1919. The L.B.O. of December 2nd 1919 carried his obituary:-

"He came into existence many many years ago when the sand trade began to develop from a purely local into a national trade........As a type he was unmistakable. He was

LEIGHTON BUZZARD SAND

generally young and hefty; a knotted muffler his neck wear; he wore his cap at a "don't care" angle, and a fag end gave him his final touch of freedom and independence. He might be met at any time of day between 7 a.m. and 4 p.m., singly or in strings, making one of his 4 journeys a day between Shenley Hill or Double Arches and Grovebury Crossing, and he had the reputation of beating all records in the twin arts of wearing out horses and roads......... Nothing disturbed him, nothing perturbed him. Old ladies might glare as he rode by, basking in the sun like some eastern potentate on a throne of golden sand. Little he cared......

The war brought him to the height of his prosperity. While lesser fry such as shop assistants and the owners of one man businesses were called up he was protected even after A men from other businesses had been ruthlessly combed out. But the greater the eminence the greater the fall. Saturday (November 29th) saw the dismissal of a big batch of carters and this week will see the sale by auction of 36 of the horses and some 40 of the carts. The cheapening of the transport costs from the pits to the railway sidings will provide other and perhaps less congenial work in the pits, but the sand carter as a local institution is gone. Mr. Marks (of Bedfordshire County Council Highways section) will not weep; the aforesaid old ladies will not weep, and those 36 horse, "all out of hard work" as the auctioneer naively put it, will not shed crocodile tears."

The sand carter eventually faded out about 1922, and the light railway reigned supreme for the next twenty-five years. Before 1914 sand carting was one of the few jobs available to 15-18 year old boys.

Sand is a soft material, easy to dig by hand. Excavators were introduced into the sand pits in the 1930's, at first to strip the overburden and later to load sand, but hand loading did not cease in Arnold's pit until the mid 1960's. PoWs were used in the way by Garsides before ceasing to use men around 1945. Men were on piece-work rates and on average a man could load 24 wagons (30 tons) daily. The sand was dug by

LEIGHTON BUZZARD SAND

cutting a horizontal groove 2 or 3 feet deep, with a pick, across the base of the face so that a whole section would slide down. When the face fell the diggers did not turn and run as the advancing sand would knock them onto their faces and bury them. Instead, they faced the avalanche, and then fell to lie on top of the sand. The finer the sand the harder it was to make it fall. The sand was screened as it was loaded by throwing it through an inclined metal sieve (like those builders still use) mounted on top of the wagon or cart. Washing the sand was introduced in the 1914-1918 war and was carried out at Grovebury sidings.

Sometimes carts were taken to the face along sleeper roads. In other cases light tramways were laid and the sand tipped by the road for reloading into carts. The horses used would haul 4 or 5 wagons on the level but only one wagon if there was a steep climb out of the pits. If, as was often the case, the wagons were not brake fitted they were stopped by jamming a stout wooden pole, 6 or 7 feet long, under the wheels.

The screens for the sand are of two main types. The shaker, with a vibrating screen of $\frac{1}{2}$" mesh made of taut piano wire, is sufficient for building sand. Sand for filter beds, foundries etc. needs to be washed to remove gravel and very fine sand or silt. A conveyor takes the sand from the tipping hopper to a sloping trough where it is washed by a stream of water onto a vibrating screen with $\frac{1}{8}$" holes called the "Niagara". The sand falls through the holes (if it were dry $\frac{1}{2}$" mesh could be the smallest that could be used) while the gravel slides off the screen and down a chute into a wagon. The sand is washed into a tank where it is stirred and the silt washed out to be pumped away to old workings. A bucket conveyor lifts the sand out of the washing tank to a chute which feeds the wagons. Perforations in the buckets allow water to drain off, and the wagons have a row of holes in their bottoms as well. Barrel washers, comprising a horizontal rotating drum made of sieves of gradually increasing coarseness, used to be used. These worked on much the same principles as above.

LEIGHTON BUZZARD SAND

At Grovebury much of the sand lies below the water table so the pits are naturally flooded. In this case the sand is excavated by a suction dredger mounted on the lake. The sand and water mixture is pumped to land through a 9" diameter iron pipe supported by small pontoons - pairs of oil drums in a simple wood frame - at intervals. On the edge of the pit the pipe discharges into elevated hoppers where the water is drained off. Further washing is unnecessary but screening is required to remove gravel and grit.

Overburden was removed by gangs of 4 men. It was removed in strips 18' wide, running perpendicular to the face. A few inches of the underlying sand would be dug out to a depth of 4' into the face, leaving three 18" square pillars of sand at the front to support it - one each end and one in the middle. One man would then begin to chop these legs while another kept watch for cracks. As soon as the section began to fall - and it was not always necessary to remove all the legs - the digger ran. Sometimes there were casualties. The loosened soil was removed by wheelbarrow or a $\frac{1}{2}$ cu.yd. truck running on very light gauge tracks to the spoil tip in a worked-out part of the pit. The barrow runs were carried across the face to the tip on tall wooden trestles. If trucks were used the track would be laid as a Y. One pair of men on one arm had a side tipping wagon and the other pair on the other arm, an end tipping wagon so that together they could lengthen and widen the spoil embankment. The wagons would be pushed by men, or sometimes a horse was used.

When excavators were introduced the overburden they dug was removed by rail, either horses or later locomotives being used. The first excavators were steam which burnt a tipping wagon load of coal a day. They had caterpillar tracks with wide wooden shoes. Up to 40' of overburden is now being removed using bulldozers and scrapers. Often it is done by contractors instead of direct labour as in the past.

The two major sand pit owners now trade as Joseph Arnold & Sons Ltd., and George Garside (Sand) Ltd.

LEIGHTON BUZZARD SAND

Joseph Arnold & Sons Ltd.

In the mid nineteenth century Joseph Arnold was a London builders merchant with a pit at Leighton Buzzard to supply himself with sand. His sand trade grew so he opened a depot in Union Street (Now Grovebury Road) Leighton Buzzard, near L.N.W.R. Grovebury sidings where the sand was loaded for dispatch. Sand was washed and screened at this depot. The 1910 Trade Directory for Bedfordshire included Joseph Arnold, Sandy Mount, Plantation Road, Leighton Buzzard, "Silver and building sand pits proprietor, peat cutter and peat and loam merchant, contractor to H.M. Government, 24, Market Square and Grovebury." He died in 1911. There was also George Arnold, sand merchant, 9, Vandyke Road, but apparently not related. In 1914 Joseph Arnold was at Union Street and Grovebury while by 1924 this had changed to Billington Road alone, probably moving soon after the opening of the L.B.L.R. in 1918.

Arnold's had quarries at Spinney Pool (c1880) and Pratt's, both in Billington Road; Rackley Hill, Grovebury; Chances (or Nine Acres) at Shenley Hill; Chamberlain's Barn (opened about 1912) close to the eastern edge of Leighton Buzzard; Twenty One Acres (opened early 1920's, closed 1930's) near Nine Acres, and New Trees, also at Shenley Hill, opened in 1963. The First World War called for greatly increased output and a new quarry under shallow cover was opened at Double Arches in 1915. All these quarries, except Spinney Pool, were subsequently connected to the L.B.L.R. At about the same time Rackley Hill quarry was sold to George Garside. There are also several small quarries in the neighbourhood not connected to the L.B.L.R.

The Union Street depot was not connected to the L.B.L.R. so new screens and washers were built on a cramped site in the northern corner of Gregory Harris's old Billington Road pit. A tipping dock was built also beside an L.N.W.R. siding south of Garside's.

Sand was also transported from the Leighton Buzzard area by canal. Arnold's had a wharf at Old Lislade (909270) and they also tipped at Branton's Wharf, Linslade (915250).

11

LEIGHTON BUZZARD SAND

Joseph Arnold & Sons Ltd. was formed in 1937. The London office was not closed until 1967.

George Garside (Sand) Ltd.

George Garside began to trade in sand c.1890. The 1910 Bedfordshire Trade Directory describes him as "silver and building sand pit owner, gravel and peat merchant, brickmaker, 28, Lake Street". By 1914 he had moved to White House, Hockliffe Street.

He had no children but was assisted by a nephew Hugh F. Delafield. After the First World War he returned to his uncle's business, although in 1920 he was given a separate entry in the Trade Directory. When George Garside died in 1926, Mr. Delafield managed the firm for Mrs. Garside until she died in 1931. Complete control then passed to him until he died in 1957. Then the business passed to his sons J.G. & W.H. Delafield, who have since entered on a period of expansion and modernisation. One of the first results of this was the formation of a limited company, George Garside (Sand) Ltd. in 1960.

Garside's first pit was in Billington Road. Part of the site was later occupied by his firm's washery. A horse worked light railway 50cm gauge (1'7½") was used to carry sand to a main line siding. This pit was worked out by about 1916, so he began to quarry at Grovebury, on the other of the LNWR Dunstable branch. He bought Rackley Hill quarry from Joseph Arnold & Sons and in addition, opened two new pits in a field nearby (behind the present Redland Tile Works.)

Rackley Hill became flooded as the quarry deepened. As the pit was nearing the end of its life a new one, the present Grovebury quarries were opened c.1926, half a mile to the south. An extensive 2' gauge railway system developed here. The Trade Recession of the late 1920's brought the need to economise. Production was then concentrated at the new Grovebury pit and towards the Watling Street.

LEIGHTON BUZZARD SAND

The war time demand for sand had made transport problems of secondary importance (all the existing pits were close to the LNWR)and so a quarry was opened at Double Arches. This was served by the LBLR when it opened in 1919. Another quarry was opened nearby in 1926 at Munday's Hill, opposite Miletree Farm, and connected to the LBLR. There were also small quarries at Heath and Reach but they did not have railways.

A washery, screening and drying plant with coal fired ovens, was built in Billington Road quarry soon after the LBLR opened. This lasted until 1964 when a large new drying plant was opened at Eastern Way, where Garside's already owned land. Until about 1946 the Company owned barges on the Grand Union Canal, but canal traffic dropped in favour of rail or especially road, and the barges were sold.

THE LEIGHTON BUZZARD LIGHT RAILWAY

THE LEIGHTON BUZZARD LIGHT RAILWAY

The first proposals to have a railway running north eastwards from Leighton Buzzard were made in about 1892 to the LNWR, but that company was not interested. Accordingly the promoters turned to the Midland Railway, and in May 1899 the Leighton Buzzard and Hitchin Light Railway applied to the Board of Trade for a Light Railway Order. The public enquiry was held on November 9th at the Assembly Room, Corn Exchange, Leighton Buzzard before the Earl of Jersey and Colonel Boughley, the Light Railway Commissioners appointed to "inquire into the expediency of granting the application made by Messrs. W.S.Cowper, John Waugh and John Henry Green for an Order to authorise the L.B. & H.L.R." The report of this inquiry was published in the "Leighton Buzzard Observer" of November 14th, 1899.

The length of the line, of standard gauge costing £98,631, was given as 19 miles 7 furlongs and half a chain. H.C. Richards, M.P., one of the two barristers appearing for the promoters, said that there were a number of sand pits close to Leighton and it was naturally desired to connect them, so as to convey the sand "a product of some value", either to the LNWR or MR. Development in the brick trade were also expected as sand and clay were to be found in the pits.

"Joseph Arnold, of Camden Town, said he was engaged in the sand trade in that neighbourhood. The line would save him 1/- per ton cartage......he carted 50,000 tons in a year. It would also enable him to complete more successfully with other markets.....George Garside, also interested in the sand trade, gave evidence to a similar effect." (Leighton Buzzard Observer Nov. 14th 1899.)

However, the Order was not confirmed, as the promoters abandoned the proposal. A new application was made in May 1902 by John Cumberland, John Waugh, Henry J. Green with a modified route. An article in the "Leighton Buzzard Observer" of June 10th, 1902 said that the required capital £100,857 had been guaranteed. "We must congratulate

THE LEIGHTON BUZZARD LIGHT RAILWAY

(the promoters) upon the prospect of the accomplishment of an undertaking which appeared to less hopeful spirits a mere dream of the enthusiasts." New industry was expected in the town from "manufacturers who are seeking to remove their works into the country."

In September the Leighton Buzzard Observer carried notices of the public enquiry to be held on Wednesday, September 24th, 1902 at the Leighton Buzzard Corn Exchange. Again the Commissioners were the Earl of Jersey and Colonel Boughley, and "there was a large company in the room".

One modification to the previous route was in the neighbourhood of Eggington where the line was to be taken to within half a mile of the village and a station provided which would be of benefit to the travelling public (268 inhabitants at Eggington) and "would have a considerable commercial influence having regard to the facilities for transferring sand, which is plentiful in the neighbourhood". Market produce would be encouraged, and the sand traffic would be a useful source of revenue. The line would begin at a junction with the L.N.W.R. Leighton-Dunstable branch approximately 2 chains south east of the bridge over the River Ouzel, and run in a generally north east direction (presumably following the Clipstone Brook for part of the way) with stations at Leighton Buzzard, Eggington, Hockliffe, Toddington, Harlington (where it would pass under the M.R. line south of the station by a brick arch bridge able to carry six tracks) Barton in the Clay, Shillington and Pirton, with possibly one at Ickleford where it joined the M.R. Bedford-Hitchin branch close to the Cadwell Road bridge. The M.R. branch met the G.N.R. about $\frac{3}{4}$ mile further on. The distance to Hitchin from Leighton Buzzard would be halved, from 41 miles to 20 miles; the L.B. & H.L.R. having a length of "19 miles, 1 furlong, 5 chains (or thereabouts.)"

There was no opposition from the three main line companies concerned and very little from the owners or occupiers of land to be used by the railway (14 dissentions out of 362 people affected.) The line would be standard gauge, steam operated, with a 14 ton axle load and 25 m.p.h.

THE LEIGHTON BUZZARD LIGHT RAILWAY

speed limit. Rail was to be at least 60 lb. per yard. Home signals interlocked with the points were required at all crossing places, and if the signals were not visible for ¼ mile then distants were also required. Platforms were required unless all coaches used had a convenient means of access from the ground but there was no **obligation** to provide shelter or conveniences at any **station stopping** place. Trains would have worked through to Leighton Buzzard LNWR station. Capital was £120,000 in £1 shares.

Although the Board of Trade gave its approval on April 6th, 1903, with five years allowed for the completion of the works, nothing appears to have been done. When the situation became aggravated following the upsurge in sand traffic during the Great War, Messrs. Arnold & Sons made attempts to have a railway built by the Ministry of Munitions but they were unable or unwilling and nothing happened. Early in 1919 the sand companies were again made responsible for the damage they caused to the roads and at the same time foreign producers were trying to re-establish the English markets. To maintain their position the sandpit owners decided to build a light railway themselves.

The plan was made public at the Annual General Meeting of the Leighton Buzzard U.D.C. on the evening of Thursday, April 17th, 1919.

"The scheme, briefly, is to construct a light tramway which will tap all of the sand pits at Two Arches (sic) and Shenley Hill and deliver sand into the railway tracks at Billington Road siding without bringing it through the town........Though at first the line will be only a single line for horse drawn traffic with loops at intervals, enough land is being taken in to provide room for a double track. The rails are heavy enough to carry any locomotive made for 2' gauge; and it is contemplated that steam will eventually replace horses as motive power. All the gradients will be easy, and one horse will be able to draw three one ton trucks on each journey............The

THE LEIGHTON BUZZARD LIGHT RAILWAY

object frankly is to retain for Leighton
Buzzard the enormous sand traffic captured
during the way by the cessation of foreign
competition." (Leighton Buzzard Observer
April 22nd 1919.)

The U.D.C. passed unanimously a resolution
recommending that the County Highways Committee approve
the level crossings specified on the plan. Joseph
Arnold, on behalf of himself and Garside and Harris,
sand quarry owners at Leighton Buzzard, applied in May
to the Committee for permission to construct 9 level
crossings where shown on the plans. These were across
Eastern Way, Shenley Hill Road, Vandyke Road, Hockliffe
Road, Stanbridge Road and three (one double track)
across Billington Road. Hearing of the approval of
the Leighton Buzzard U.D.C., "the clerk of the Council
was instructed to prepare an agreement with Messrs.
Arnold & Sons for the construction of such level crossings as may be found necessary, and to enter into the
agreement on behalf of the Council."

On June 3rd, 1919, a meeting was held at the
Swan Hotel, Leighton Buzzard, to hear details of the
proposed light railway to carry sand. J.H.Green was
elected Chairman - "he knew nothing about the railway,
except it was in the air" (!) Also present and responsible for the scheme were Messrs. A. & E. Arnold,
H.W. Clough, H. Delafield, A.W.Merry, R.G.Walton;
G. Garside was unavoidably absent. Mr. A. Arnold
explained the difficulties of the sand trade and said
that the only solution lay in the construction of a
railway which might also carry farm produce in special
wagons.

Mr. Walton, who had been interested in the railway from the start gave details of the proposed Company,
the Leighton Buzzard Light Railway Limited. The nominal
capital would be £20,000 in £1 shares and the estimated
capital expenditure was £15,000. The sand merchants
had guaranteed a minimum annual traffic of 70,000 cu.yds.
for 10 years, the rates for which were open to periodic
revision. With an estimated annual expenditure of £2875
including sinking fund he thought that a 10% dividend was

THE LEIGHTON BUZZARD LIGHT RAILWAY

probable. Wayleaves over most of the route had already been obtained for 30 years and few difficulties were expected in completing this part. The directors of the proposed company were A. and E. Arnold, G. Garside, H. Delafield (sand merchants) R.G.Walton (Solicitor) T. Bromhead Bassett (backer). Most of the capital had been promised at the meeting and the remainder soon subscribed, the sand merchants contributing about one third.

The Company was "incorporated the 26th day of July 1919", with its Registered Office at 20 Bridge Street, Leighton Buzzard. The objects for which the company was established were to construct, purchase, or lease light railways in Bedfordshire and Buckinghamshire and to fit out, maintain and operate the same with "horse, or by electricity, steam or other mechanical power". The Company could provide road services for passenger and freight, could generate and supply electricity for any purpose, and could carry on practically any business except that of sand merchants.

The Company was a "Private Company" within the meaning of the Companies Act 1908 and 1913. This meant that it:-

(a) restricted the right to transfer its shares.
(b) limited the number of shareholders to 50.
(c) prohibited <u>any</u> invitation to the public to subscribe for its shares and debentures.

In return for these restrictions the Company gained a number of privileges, including the right to commence trading as soon as it was incorporated and (until the 1967 Companies Act) was freed from the necessity of sending a copy of each year's accounts to the Registrar of Companies, Bush House, London. The Company while having the advantages of limited liability, was thus enabled to maintain considerable privacy in its affairs. The L.B.L.R. has in fact deposited very few records with the Registrar.

Construction of the railway was soon started and proceeded rapidly as it had very few earthworks and those of a minor nature. The consulting engineers were Babtie, Shaw & Morton of Glasgow, using the detailed scheme prepared by Mr. H.F.Firth. The contractors were Messrs.Lamb and Phillips of 107, Clerkenwell Road, London, E.C.1.

THE LEIGHTON BUZZARD LIGHT RAILWAY

They appear to have skimped the work by laying 25 lb instead of 30 lb rail, and by springing rails for curves instead of bending them properly. Under traffic curves straightened and had to be relaid.

The formal opening of the L.B.L.R. took place on the afternoon of Thursday, November 20th, 1919, although the branch lines to the pits remained to be completed. The ceremony began with lunch at the Swan Hotel, attended by the directors, some shareholders and others, including Mr. M.G.Townley, M.P., "who had left the Commons in spite of a three line Government Whip because he felt his duty was with his constituents, especially in regard to such an undertaking". Subsequently Mr. Townley "baptized the two locomotives in traditional manner and the guests then traversed the line in a specially constructed temporary train, made of forms fastened to four trolleys, and drawn by a petrol engine (the contractor's) the locomotives being too smoky for passenger traffic. On some parts of the line a good speed was obtained - probably nearer 20 mph than 10, and some of the inexperienced passengers reflected uneasily of the 2' gauge beneath them. But all passed off well and when the wind made sport of hats the train obligingly stopped to permit their being recovered. Despite the absence of springs on the temporary train there was little jolting, a fact which testified eloquently to the solidity of the line.....At the far end of the line the party gathered in a building and speeches of congratulations were made. Mr. Townley said he had been approached by Mr. Arnold and others as to the desirability of building a light railway and it reflected great credit....that the work had been executed so quickly.....Mr. Marks, Clerk of the Bedfordshire County Council promised to help and carried it out to the letter. The line was the first light railway constructed since the war. It would have been to the country's advantage if it had been built five years earlier....Some of the land passed through looked like market gardening land, which would benefit, and the railway could bring in manure. He congratulated the directors, and the L.N.W.R. for bringing in material so promptly. He wished it every success."
(Leighton Buzzard Observer, November 25th, 1919).

THE LEIGHTON BUZZARD LIGHT RAILWAY

    Mr. Walton said that there had been no obstruction to their endeavours, they had built 4¼ miles of railway on wayleaves "without having to purchase a single piece of land or invoking the assistance of the Light Railway Commissioners." Telegrams of congratulation were read from the Refractories Association"..... and congratulate the directors on the object lesson they are giving the Ministry of Supply" and Mr.Syme of the Ministry of Munitions who telegraphed good wishes and apologised for not attending. The meeting finished with a vote of thanks for Mr. Townley proposed by Mr. J.H.Green.

    The railway was soon completed. The majority of the sand carters were discharged as redundant on Saturday November 29th and rail traffic presumably began the following Monday, December 1st. It may be coincidence, or a reflection of the date of opening for traffic, but the L.B.L.R. financial year ends on December 2nd. In the issues of November 25th and December 2nd of the Leighton Buzzard Observer, there appeared advertisements for an "Important Sale of Contractors and Hauliers Stock", the Sale being due "to completion of the new Light Railway". The Sale was to take place on December 3rd at Arnold's Union Street depot.

    This sale was on behalf of Messrs. Joseph Arnold & Sons and George Garside. Included were :- 36 horses, "all out of hard work", 40 constructors carts, 3 large sand washers by "Baxter" of Leeds (from Arnold's Union Street depot?) "about 500 yards Jubilee track 20"" followed by "12 tip wagons, points and crossings" presumably from internal sand pit lines now linked to the L.B.L.R. and, surprisingly, "6 contractors tip wagons 3' gauge," possibly from a sand pit but otherwise of unknown origin.

    One curious feature is the lack of any mention of the L.B.L.R. in the Ministry of Transport (previously Board of Trade) records. A normal light railway owned its trackbed and the L.B. & H.L.R. had compulsory purchase powers, but the L.B.L.R. is laid entirely on rented land for which

THE LEIGHTON BUZZARD LIGHT RAILWAY

wayleaves are paid. Furthermore, if a light railway wishes to alter its name, route or gauge it needs a Ministry Order to grant the necessary authority. No such amending order was made for the L.B. & H.L.R. and Mr. Walton (see above) said they did not apply for any Light Railway order. It would appear therefore, that despite its title the L.B.L.R. is not a proper light railway (unfortunately there is no legal definition of the term) but is in the same category as other industrial railways such as the lines in the Midlands ironstone field. There is however, the differences that the L.B.L.R. itself does not extend into the actual quarries - it is purely a carrier.

LATER HISTORY

LATER HISTORY

Traffic steadily increased over the years; the L.B.L.R. supplying motive power and maintaining the track, while the quarry owners provided their own wagons. The usual dividend paid until the early 1950's was 15%, fluctuating to 17½% and 12½% in good and bad years. The line's heyday was in the period between 1934 and 1939, and the years following 1945 when petrol restrictions discouraged road traffic and the demand for sand boomed. It was necessary to lay double track between Stonehenge and Munday's Hill in the late 1940's. Arnold's dispatched up to 300 wagons daily and Garside's 200, making some twenty train loads. In 1948 the Company purchased the freehold of their depot at Billington Road and built a new locomotive shed and workshops at a cost of £3018, completed in 1950. This, and a length named Miletree Furlong, was the only land the L.B.L.R. owned. (A piece near Leedon was sold after 1945). Wayleaves costing about £550 p.a., accounted for almost all of the route. Between 1950 and 1954 three new locomotives were purchased, and soon after three more, secondhand.

From 27th May, 1948, the Registered Office was moved to 6, Church Square, Leighton Buzzard. During the war all sign posts, place names etc. had been removed to confuse invaders. The company plate was therefore amended to read "Light Railway Company Limited", to the subsequent confusion of one enthusiast at least. !

As postwar restrictions on road transport were eased, decreasing quantities of sand were sent away by rail. This trend was accelerated when British Railways suffered from a 17 day strike in June 1955. This caused many of Arnold's and Garside's customers to have their sand delivered by road, loaded in many cases - especially Garside's - at the quarry. L.B.L.R. traffic fell disasterously, and the dividend fell to below 10%. A reorganisation was put into operation on December 3rd, 1958.

From this time the L.B.L.R. would own and maintain the track and the quarry companies haul their own

LATER HISTORY

traffic, paying tolls - 2/4½ per wagon in 1968 - to the L.B.L.R. The ex-W.D.L.R. locomotives which had served the line well for nearly 40 years were all withdrawn for scrap and the six modern ones sold - four to Arnold's and two to Garside's, as this was roughly the proportion of traffic handled. Following the change, there were usually two of Arnold's large locos. and one of Garside's each making about 3 journeys a day, the last leaving Double Arches around 3 p.m.

As the founders died so their shares were divided, often passing to people with no connection with the sand trade. The rights of the non sand-pit owner shareholders were safeguarded by a special resolution passed on February 2nd, 1948, which amended the articles to provide for a board of six directors, three to represent sand merchant share-holders and three to represent the remainder. The sand merchant directors had to be customers for the time being of the railway. In 1950 there were 38 shareholders, with holdings ranging from 18 to 1701 shares. The issued capital in 1959 was £15100, for many years it had remained at the £15000 originally issued.

A special resolution on July 27th, 1949, gave the company power to distribute to its members any surplus from the realisation of capital assets (as distinct from revenue or business profits) and defined such surplus as meaning monies over and above a sufficiency of assets to cover all liabilities including paid up capital. This confirms that the company had been quite profitable.

The L.B.L.R. had bought investments in addition to distributing its profits. In 1942 it had £3300 in defence bonds and a building society. In 1947 this had risen to £5200, and by 1953 there was £6500 in the building society. Interest from these investments helped maintain the dividend. The money realised by the sale of its locomotives and workshops was invested too, so that following 1958 most of the L.B.L.R. dividend was obtained from

LATER HISTORY

these investments, not from its railway operation.
The Company's accountants were Keens, Shay, Keens &
Co. of 14, Church Square, Leighton Buzzard.

   Garside's traffic continued to fall, so that
the L.B.L.R. became increasingly Arnold's railway.
On January 1st, 1963, Arnold's purchased all L.B.L.R.
shares at par and the railway became a subsidiary, its
registered office now being Arnold's office. Such
Garside's traffic as used the railway was still hauled
by Garside's locomotives but tolls were paid to Arnold's.
Near Miletree Farm the L.B.L.R. runs on Garside's land,
so Arnold's pay them the way leave, while in their turn
Garside's pay tolls to Arnold's for their trains.

   The old sand plants at Billington Road were
becoming less and less economic to run. Garside's
built a large modern drying and grading plant at Eastern
Way, which came into use in 1964. Their traffic to
Billington Road ceased completely by the end of the year,
but trains still shuttle continuously between Munday's
Hill and Double Arches quarries and the new plant, from
which all sand is dispatched by lorry.

   Arnold's built a new washery at Double Arches
in 1963. Water was readily available and the new plant
could be built without interfering with the existing
equipment. Furthermore, the Billington Road site was
cramped and it was easier to dispose of reject sand in
the pit than to have to haul it back to Double Arches.
Following this, Arnold's main line traffic fell, the
only sand taken to Billington Road being for dispatch
by rail. One large locomotive making 3 daily journeys
handled the traffic.

   Following various reappraisals and modernisation of methods, British Railways increasingly gave the
impression of being interested solely in train load
traffic. Wagon load sand traffic has steadily diminished and, as a consequence, the tonnage hauled to
Billington Road dropped so that now only a single train

LATER HISTORY

sufficed usually, and that did not run every day. Early in 1969 British Railways announced that they wanted to close the remnants of the Dunstable branch serving Billington Road and Grovebury sidings as it lost £26,000 p.a. Closure will probably take place in December 1969.

Since after the closure of Billington Road sidings the only light railway traffic south of Vandyke Road would be locomotives travelling to and from the workshops it would not be economic to retain the whole railway. The enthusiast group, the Iron Horse Preservation Society, established in 1967, therefore has agreed to take complete responsibility for this section, including wayleaves and rates, as from 1st May, 1969. The Society will then have the sole use of the section and Arnold's locomotives etc. will go to and from the workshops by road. The future of the section to Stonehenge is also uncertain. At present it is used by trains from New Trees quarry taking sand to Double Arches for washing, but would be uneconomic to retain it in the absence of any Billington Road traffic. Arnold's have therefore applied to the County Council for permission to use lorries instead, running along Shenley Hill Road, and the railway is used in the meantime.

Beyond Stonehenge the future of the railway is secure as the brickworks takes a lot of sand from Arnold's Double Arches quarry and the section beyond Munday's Hill carries heavy traffic for Garside's as well. This is however internal traffic, and once again the output of the sand pits is put onto the local roads.

The whole line may survive if the Iron Horse Preservation Society achieves its aim of running regular passenger services over the L.B.L.R. The existing wayleaves expire in 1979 and some landowners are believed to be unwilling to renew them when the time comes. This date may mark the complete closure of the line, or of large parts of it. Much depends on the use the quarry owners make of the northern section - many quarry railways have succumbed to dumper trucks and /or conveyors in recent years - and on the success or otherwise of the Iron Horse group.

THE DESCRIPTION OF THE ROUTE

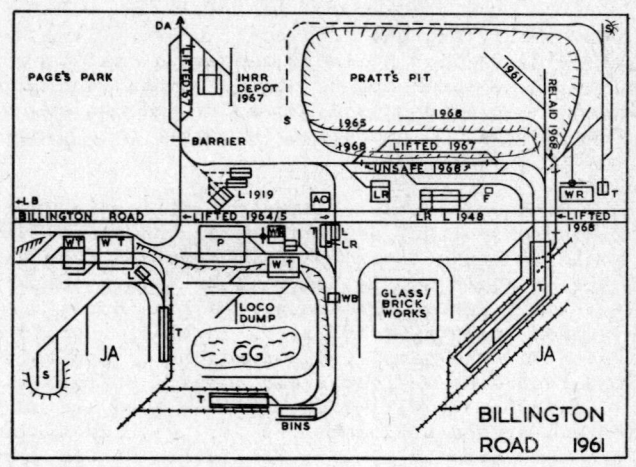

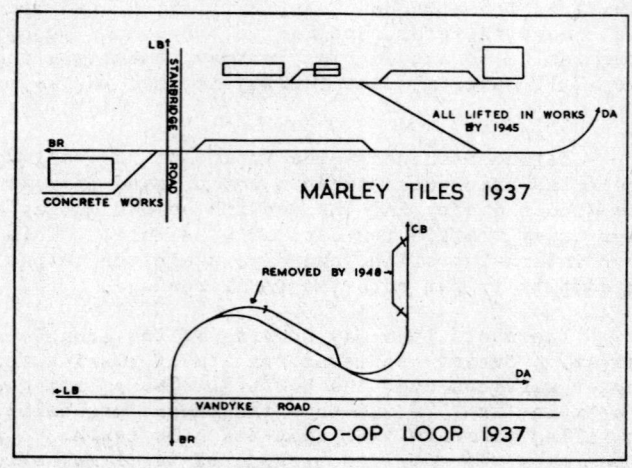

THE DESCRIPTION OF THE ROUTE

THE DESCRIPTION OF THE ROUTE

The Leighton Buzzard Light Railway begins (or bearing in mind the main traffic flow ends) on the eastern side of Billington Road (929240) about $\frac{3}{4}$ mile south east of the centre of Leighton Buzzard and a few yards north of the Dunstable branch crossing. The layout has not altered a great deal over the years.

It begins at the southernmost, and now sole remaining, of the three ungated level crossings over Billington Road. The company was responsible for these crossings. Three tracks (two in 1926, and since mid 1968) cross the road giving access to a gantry for loading main line wagons and a former glassworks west of the road. East of the crossing the centre line divides, one branch to join the southernmost track and descends into Arnold's Pratt's Pit. The other branch, and northernmost crossing track turn sharply northwards and run parallel to the road, soon dividing to form three loops each about 100 yards long. Long sidings make trailing connections with each of the outer loops. The eastern one was latterly out of use owing to the proximity of Pratt's Pit. The other, between the loops and the road, gives access to the L.B.L.R. locomotive shed.

This shed was built in 1948/49. Constructed out of bricks with steel framing it measures 40' by 110', the last 25' being the workshop. While there are three pairs of doors in the southern end only two tracks are laid. The shed can hold about 20 locomotives. At its inner end is a small but well equipped workshop. When the L.B.L.R. ceased to provide motive power in 1958 this shed was taken over by Arnold's to house their main line locomotives, Billington Road shunters and any spare quarry locomotives. All Arnold's locomotives are now repaired here. Previously Arnold's locomotives had been kept and repaired in a corrugated iron shed immediately north of the 1948 building, by the end of the loops.

DESCRIPTION OF THE ROUTE

The single track now begins to climb and the branch to Garside's washery made a trailing connection. The gradient steepens to a maximum of 1 in 20 and the line turns sharply to the east to run beside Page's Park. At the top of the gradient and end of the curve, the steeply graded branch (about 1 in 20) to Arnold's washery made a trailing connection.

In the angle between the two lines stood the original L.B.L.R. locomotive sheds, long disused but not pulled down until about 1966. The three buildings, two single road and one three road, with a pit on the centre track only, had all been built by 1926. They measured 36' by 13', and 33' by 14' (these have been demolished) and 18'6" plus 2' extension by 12'. The central one was presumably the original steam shed, holding one locomotive (or two 20HP petrol locomotives after the extension had been made,) as the 33' long shed was originally 22' long.

Just beyond the junction a lifting barrier made from a length of rail and pivoted about 18" above ground level was erected in the early 1960's. It can be padlocked horizontally and was to prevent children pushing wagons out of Page's Park loops. On the level ground beside the park is a loop about 100 yards long like all the others on the line, on the left (northern) side. Following complaints that Garside's trains waited here excessively long times, blocking the line, a second similar loop was laid on the right hand side of the line some time after 1926. In 1967 this loop was lifted and the further point used to give access to the Iron Horse Preservation Society shed and sidings.

The line curves slightly northwards and then runs more or less straight and between hedges at the edge of fields, climbing to a summit at Red Barn loop (660 yards). This loop was laid sometime between 1926 and 1937. The line falls at 1 in 60 past a housing estate - married quarters for R.A.F. Stanbridge - and to link this estate to the main part of the Station a steel footbridge with approach ramps was built in 1962 at the foot of the bank. The track then levels for the 200 yards to the Stanbridge

DESCRIPTION OF THE ROUTE

Road crossing ($\frac{3}{4}$ mile). On the right, immediately before the crossing is the Leighton Buzzard Concrete Co.Ltd. works whose siding makes a trailing connection. This had been laid by 1926 and rail traffic ceased 10th September 1968. Most of the siding is laid in bridge rail. Loaded wagons were uncoupled from the main train at the nearby loop (or at Leedon loop if the train was divided for the ascent) and pushed into the works. After the crossing the line begins to fall (1 in 80) as it passes through the first Marley Tile Works (opened in 1930). There was a loop here by 1926. Several sidings were laid into the works soon after it opened, but they disappeared around 1940 and sand delivered by lorry. Up to 140 wagons daily used to be sent here from Double Arches. Beyond the loop the gradient steepens for 220 yards, averaging about 1 in 30 with a maximum of about 1 in 25 near thefoot. There is a cutting about three feet deep on the centre section of the bank and then, near the foot, the line is crossed by a concrete road linking the two halves of the Marley Tile factory. Older crossings, since abandoned, are nearby. There are a large numbers of foreign workers in Leighton Buzzard so these crossings are protected by notices - "Beware of the trains. On no account is the line to be obstructed" - in four languages - English, German, Jugoslav and Polish. The new crossing also has mirrors beside it to show approaching trains, because the railway is now invisible from the road.

At the foot of the bank the line crosses a small stream (1 mile) and turns sharply with a check-railed curve to run slightly west of north, more or less level beside a hedge. After about 300 yards there is a second small bridge and a very short siding on the right entered by a facing point. This is the "dud loop", and except when a train is passing, the points are kept facing the siding to deflect runaways into the field. Leedon loop follows, crossed by a footpath and the line reaches **Hockliffe** Road (A4012) crossing ($1\frac{1}{4}$ miles) at Leedon.

DESCRIPTION OF THE ROUTE

Still falling gently and running along the edge of fields, the railway continues northwards, before running across the centre of a field on a low embankment which curves round to the northwest and brings the line to its major engineering feature, the Swing Swang bridge over the Clipstone Brook ($1\frac{5}{8}$ miles). This is a single girder of 15' span, with brick abutments and iron railings. A straight climb of about 1 in 50 brings the line to the Vandyke Road. The level crossing ($1\frac{3}{4}$ miles) is followed immediately by a long right hand curve of about 70' radius leading into a shorter left hand curve to bring the railway beside the road, and separated from it by a hedge, a position it keeps for the remainder of its length. This curve is known as the Co-op loop (the Co-op used to own the land on which the nearby school is built.) In 1926 and 1937 there was a loop here, but since the last war this had degenerated into a long trailing siding, mostly overgrown. By the siding points, the branch to Arnold's Chamberlain's Barn quarries makes a facing connection and curves sharply away to the left. The L.B.L.R. boundary is just before the points where the Chamberlain Barn and New Trees lines divide, about 300 yards from the main line.

The track climbs steadily for about 700 yards. Close to the summit is Bryan's loop and a PW hut. A dip and rise occur, followed by a descent to Shenley Hill Road crossing ($2\frac{5}{8}$ miles). The descent continues for about 200 yards more to the Driroof Tile works and Stonehenge Brickworks ($2\frac{3}{4}$ miles) behind which is Arnold's, Nine Acre (or Chances) Quarry. A long climb follows, with a maximum gradient of about 1 in 27. This section was laid with double track around 1945, but by the mid 1960's the western loop line had fallen into disuse following the reduction of traffic. The summit is by Mile Tree Farm and there is a descent to Garside's Munday's Hill Quarry ($3\frac{1}{4}$ miles) which has a loop. The quarry branch used to leave from this loop, but since 1966 it has made direct connection with the main line a few yards further on.

DESCRIPTION OF THE ROUTE

The remainder of the route is more or less level. About 300 yards further on there is a triangular junction, laid in 1964 to serve Garside's new screening and drying plant about 200 yards north of the line. Another 100 yards and a facing connection is made on the left. This track immediately divides, one line to run parallel to the main line but climbing to reach a wood and steel tipping dock, disused since about 1945. The other line continues straight to a corrugated iron building, formerly Garside's locomotive shed, and later workshops for repairing their excavators and other plant. Since 1964 locomotive repairs have been done here too.

From the workshop junction a further 100 yards brings the line to Eastern Way crossing (3½ miles). A few yards beyond the crossing is a pair of points, owned by the L.B.L.R. and marking the boundary. The left hand branch leads into Arnold's Double Arches quarry yard while the right hand branch curves, crosses a stream by a sleeper bridge, passes through a hedge and enters Garside's Double Arches quarry yard. This stream passes under the road in twin tunnels and so gives Double Arches its name. The total length of L.B.L.R. main line is slightly over 3½ miles. Measurement of the 6" map gives a length of 6300 yards, while a survey my brother and I made in 1961 with a length of cod line supposedly 100' long, gave a length of 6408 yards. The contractors were paid for $8193^2/3$ yards (sic!) - a little over 4½ miles - but this included loops etc. However, a mile total for these is not excessive. The sign post at the junction of Mile Tree Road and Eastern Way gives the distance to Leighton Buzzard as 2½ miles.

Track

The L.B.L.R. was originally laid with 25lb rails although 30lb material had been agreed. The sleepers were cut locally and were of poor quality. The line was ballasted with ash. The finished track, $8153^2/3$ yards, cost £12030-13-8d. The Final Certificate was issued by the Consulting Engineers on March 10th, 1920. Latterly

DESCRIPTION OF THE ROUTE

Track (cont.)

30lb rail was used, spiked to halves of ex BR sleepers at 2'6" to 3' centres. Ballast is not now visible being hidden under sand droppings and grass. At first six men were employed on track maintenance. Around 1950 the number fell to four, then to two by 1960, to one in 1966 and in 1968 to none. Repairs were then carried out by other staff as necessary or when hard frosts stopped work in the pits. After 1958 derailments had to be reported to the L.B.L.R. so that the PW gangs could be informed. In recent years the condition of the track has deteriorated and leaves much to be desired in some sectors. All locomotives carried a pair of V shaped rerailing ramps which were sufficient for minor derailments.

In the quarries 20lb rail is the rule, laid on steel or wood sleepers. The old horse and hand worked lines used lighter rail still, about 10lb per yard. A length of 50cm (1'7½") gauge track survives. The rails are 4 metres (16'6") long and spaced by five steel sleepers.

OPERATION

OPERATION

The heavy traffic carried by the L.B.L.R. was handled with the minimum of regulation. There was only one signal, and no normal method of single line working was used. Trains ran as required, and a sharp lookout was kept for traffic in the other direction. The line ran through fairly open country, so trains could usually be seen or heard some way off. Speeds were low, so this was not quite as hazardous as it sounds, but collisions were not unknown ! If two trains met between loops, the empties had to give way and reverse to the next loop, although if the loaded train had just passed a loop it might reverse. Between Chamberlain's Barn and Stonehenge two loaded trains could meet, in which case the Chamberlains Barn one had to give way. Trains kept to the left at loops when they crossed, and the loaded train had the straight track; when not crossing all trains used the straight road.

The normal load handled by one of the large locomotives was 24 loaded wagons (about 30 tons of sand) while the small locomotives, L.B.L.R. and quarry-owned, took 10 or 12 wagons. Six ton bogie open wagons were used to carry sand in bags from Arnold's drying and grading plant at DoubleArches to Billington Road. These were taken as "equal to four" loaded sand wagons and were always marshalled next to the locomotive. A round trip tool $1\frac{1}{4}$ hours (minimum).

The railway handled a considerable traffic. During 1934-1940 and 1945-early 1950s Arnold's dispatched 300 wagons daily and Garside's 100-150 i.e. over 100,000 tons annually. Even in the late 1950's it was carrying 250 tons daily (60,000 tons yearly) and in 1961, three big locomotives (two Arnold and one Garside) were in daily use. In mid 1968 there was usually one or sometimes two trains from Double Arches to Billington Road, and even these became less frequent by the end of the year.

OPERATION

The working was organised in "trips" of 72 wagons, and normally there were four trips daily. This required a minimum of three 40 HP locomotives (or the equivalent in 20HP). Unfortunately, a 20HP petrol locomotive could only manage 10 wagons, so an extra journey was necessary. The small locomotives were also used for light winter work and short runs such as from Chamberlains Barn to Stonehenge. Locomotives made three or four round trips daily. At its busiest, 4½ trips daily were tried, and also making trips of 80 wagons, but this was unsuccessful due to the shortage of empty wagons in the quarries in the mornings, and too many for unloading at Billington Road in the evenings. When Marley Tiles received sand by the L.B.L.R. it took up to 140 wagons daily, almost half the total traffic. The Concrete Works had only 6 or less wagons daily. When the empties were collected the train was left on the main line just cleat of the points. Full wagons were marshalled at the rear of Billington Road trains. These stopped in the Marley Tile loop and the wagons pushed to the works. If the train had been divided for the ascent, then the wagons were pushed up from Leedon loop, or else they would be in the middle of the train when it re-formed.

In general the gradients favoured loaded trains with the exception of the steep bank, maximum about 1 in 25, through the Marley Tile works. All trains were supposed to have rear assistance, or divide, to reduce the strain on the couplings, but certainly after the quarry companies took over operation this rule does not appear to have been observed strictly, although if a driver is caught ascending unassisted he may be reprimanded. The practice is said to stretch the locomotive driving chains. Trains were divided at Leedon loop and half was hauled up to the Stanbridge Road loop, where the wagons were left while the locomotive returned for the remainder. If, as was commonly the case, there

OPERATION

were several loaded trains following each other, the first was banked by the locomotive off the second, its train being left in Leedon loop. By the time the second locomotive returned to collect its train, the next loaded train was about to arrive and the cycle was repeated. A variation of this was to have a locomotive on banking duty only. When banking was discontinued the last wagon of Garside's trains was fitted with a sprag - a short length of old rail fastened loosely in the coupling pocket - which trailed behind, bouncing over the sleepers. If a coupling broke, the sprag dug into the ground and stopped the train before it could run far.

A five m.p.h. speed limit was observed past Stonehenge to keep trains under control down the bank from Monday's Hill. Further precautions had to be taken at Billington Road where the level crossings into the quarry depots were all approached by steep gradients. Loaded trains stopped in the loops beside Page's Park - Arnold's trains on the line nearest to the Park; Garside's on the side furthest away. The main line locomotives ran to the back and waited until a 20hp quarry locomotive had arrived and coupled on to the front. Then, with a locomotive at each end, the train was taken slowly down the gradient into the depot. Trains of empties from Arnold's were banked up to the Park by a 20hp locomotive. Following the closure of the washeries, when only the southern tipping dock was in use, Arnold's trains were double headed down the bank from the Park to the yard by the locomotive shed by the Billington Road shunter.

Trains could be seen from Billington Road as they approached Page's Park loops. To cut waiting time there to a minimum, the 40hp locomotives had different shapes painted on their cabs for easy identification from a distance. Since any day a given locomotive hauled trains of only one company, the ownership of any train load could be quickly determined and the appropriate quarry locomotive sent to meet it.

OPERATION

This scheme appears to have been introduced in the late 1940's/early 1950's, and lasted until about 1960, by which time traffic had fallen considerably so delays did not matter so much. Usually the L.B.L.R. had 4 locomotives on Arnold's traffic and 2 on Garside's.

LOCOMOTIVE CODES

L.B.L.R. No.	A/G No.	CODE
(1)		None; but locomotive a distinctive shape.
2		White horizontal strip across end; ends of cab-front sheet white.
3		Number above white horizontal strip across end.
4		Number below white horizontal strip across end. White circle either side of vizor.
5		White central panel to cab-front sheet and a white square in the centre of the end.
6		White horizontal strip across end, and white cab-front sheet.
10	G10	None; but locomotive a distinctive shape until 11 et sq. bought.
(11)	A43	White rectangle on upper half of cab side.
12	G12	White triangle on cab side.
13	A44	White rectangle with circle cut out on upper half of cab side.
14	A42	White X on upper half of cab side.

( ) Number not carried.

     The 40hp locomotives were used for main line **traffic.** The 20hp ones for light loads, collecting wagons,

OPERATION

and for hauling sand between quarries - e.g. Chamberlain's Barn to Stonehenge. But the traffic between Double Arches and Stonehenge has always been hauled by Arnold's locomotives.

Main line locomotives carried a crew of two, a driver and an assistant, often a boy. The boy coupled up the train, changed points and flagged the train over the ungated level crossings which have gates across the railway only. He rode in or on the front of the locomotive. At crossings he ran ahead with his flag, or red oil lamp on winter mornings to open the gates. After the train had crossed the road it stopped to allow the boy to regain the locomotive. In L.B.L.R. days, if a locomotive was at a quarry waiting for a load at lunch time the crew were collected by the company car and taken to Billington Road where there was a rest room. Crews tended to keep to the same locomotive.

Many quarry locomotives operate across Eastern Way and they do not carry a second man, so since 1965 a flagman has been permanently on duty here. The sole L.B.L.R. signal - a fishtail arm mounted on a rough, unpainted 30' pole - stood beside this crossing. It was visible from the end of the double track near Munday's Hill, and combined the duties of point indicator for the two quarry branches with that of a warning of approaching loaded trains. The line blocked indication was the arm pointing straight up. The signal was not greatly used, and was derelict by 1961. In 1954 the arm was divided into five approximately square panels, three red and two white, but later the arm was completely red.

ACCIDENTS

Most railways have accidents, and the list which follows, while not complete, gives an indication of contretemps which have befallen the L.B.L.R.

At an early date a 40hp locomotive met a 20hp one, LR 2478, head on. The impact bowed the frame of

OPERATION

the smaller locomotive upwards slightly and thereafter it needed special brake blocks. Around 1923 it was sold to Joseph Arnold & Sons, but it was not scrapped until the 1950's. There is the suggestion of another collision by the order of one six-ton locomotive buffer in June, 1928. In May, 1965, a train from New Trees to Double Arches met a loaded train from Double Arches to Billington Road near Abraham's farm. The 6 ton main line locomotive was unharmed but No. 34 on the New Trees train bent its chassis and was scrapped.

Marley Tile bank has been the site of several runaways. Once a coupling broke and the train ran as far as the Hockliffe Road crossing before derailing and crashing into a cottage adjoining the line, luckily without loss of life. Subsequently the cottage was purchased by the railway and demolished, apart from a few buildings which served as a linesman's hut. Also, a "dud loop" was laid just beyond the foot of the bank. The point here was always left set for the siding to deflect any runaways into the field.

Once, the crew of a following light engine heard some runaways careering down the hill, although the light was not sufficient to see them, but they were able to jump clear before the collision which left wagons piled high on the locomotive.

In August, 1962, Garside No. 10 with a loaded train met one of Arnold's locomotives on a train of empties at the blind corner on the bridge at the foot of the bank. Arnold's driver, followed by three wagons, jumped into the brook, while Garside's driver sat tight. There was no personal injury but many wagons were destroyed. A similar accident happened at the same spot in 1964 when Arnold No.43 on a loaded train met a _loaded_ train of Garside's, who were then working sand away from their Billington Road depot before it closed down. No one was hurt, and little damage was done, but a little while later No. 43 suffered from a broken crankshaft, possibly as a result of the impact.

OPERATION

A double break away occurred on January 9th, 1962. While ascending the bank a coupling in a Garside train broke. The sprag derailed the last three or four wagons before they ran far. As the train continued up the bank the locomotive coupling snapped after a jerk and the rest of the train ran back into the derailed wagons.

Level crossing collisions with road vehicles occur from time to time. At Hockliffe Road crossing a van was rammed by a train in February 1958. The van driver was injured, but the locomotive was undamaged. In 1960, one foggy morning, a lorry collided with a locomotive here, while in 1961 a car hit a train in the early morning, the flagman's oil lamp having gone out. In 1965 a train of Arnold's empties, contrary to regulations, was being propelled by a 20hp locomotive over Eastern Way crossing before there was a permanent flagman, when it hit a car. Also in 1965 at 3.45 one Sunday morning, a car ran into a locomotive at Eastern Way. On November 28th, 1968, a train from New Trees collided with a Stonehenge employee's car at the brickworks entrance. The car driver admitted he failed to look for a train before crossing the line.

The only brakes on a train are on the locomotive. When these failed while a loaded Arnold's train of 16 tipping wagons and 2 bogie opens was descending from Munday's Hill, the train ran out of control to derail on the points at the entrance to Stonehenge.

A fatality occurred in 1951 at Marley Tile loop. A boy fell and was run over while he was taking out the pin to uncouple two locomotives.

When blocked by a heavy fall of snow the line is cleared by hand, with some assistance from Garside's snowplough. Arnold's and Garside's paid £28 and £12 respectively for L.B.L.R. snow clearance in February, 1940. In the 1962/63 winter, Arnold's trains ran with a locomotive at each end, while Garside's double headed their trains.

LIGHT RAILWAY LOCOMOTIVES

LIGHT RAILWAY LOCOMOTIVES

The L.B.L.R. locomotive stock at its opening comprised two 0-6-0 well-tank locomotives built by Hudswell, Clarke & Co.Ltd., Leeds, works numbers 1377 and 1378. They had been part of an order of 20 placed in June by Robert Hudson Ltd., Leeds, light railway engineers for the War Office, for use on forward railways in France and other theatres of war. (Hudson's supplied many firms with locomotives but always sub-contracted their locomotive orders.) These two 60 cm (1'11$\frac{5}{8}$") gauge locomotives were numbered 3207 and 3208 respectively in the War Department Light Railway List. At the end of the war they were left on the makers hands. Later they were regauged to 2' and sold to the L.B.L.R. for £1000 each, leaving the works on May 31st, 1919, to be delivered to J.Arnold & Sons, Billington Sidings.

The locomotives were a standard Hudswell Clarke design, Ganges or "G" class, supplied to a number of contractors and others since about 1910. They were of robust and simple construction. The six coupled wheels, 1'11" in diameter, were closely spaced to give a rigid wheelbase of 4'2" which with the centre pair flangeless, enabled the locomotive to traverse a curve of only 35' radius. Two outside cylinders, 6$\frac{1}{2}$" in diameter by 12" stroke, drove the rear coupled wheels. Outside Walschaert's valve gear was fitted. The water tank was between the frames and held 110 gallons. This and the lack of foot plating caused the bouler to appear very high pitched. The boiler barrel, 5'6" long, 2'1" in diameter, made of $\frac{3}{8}$" steel plates had 45 1$\frac{5}{8}$" OD steel tubes and supplied steam at 180 psi. The firebox was of best $\frac{3}{8}$" copper plate, the crown and sides being one piece. Heating surface was - Firebox 18.3 sq.ft., tubes 109.2 sq.ft. total 127.5 sq.ft.; grate area 3.75 sq.ft. The dome was fitted with a pair of Pop type safety valves, and a sandbox was also mounted on top of the boiler. Side bunkers held 16 cu.ft. of fuel. The "driver's shelter" was open above the waist except at the front. The locomotives were 15'5$\frac{1}{2}$" long over buffers

LIGHT RAILWAY LOCOMOTIVES

5'8" wide over cab. 8'6" high to the top of the chimney. In working order they weighed 6 tons 17 cwt. The tractive effort at 75% boiler pressure was 2970lbs.

At Leighton Buzzard the two steam locomotives were painted black and do not appear to have been numbered. They did not last long, however, various reasons being given by old employees. It has been said that they were heavy on repairs, being especially liable to steam pipe fracture and to the abrasive effects of sand in the axle-boxes and valve gear. Then they were said to have been unsteady, often rolling off the track, or spreading the rails on curves. This was attributed to their weight, but the track had not been that well laid in the first place. Their lack of water capacity was another alleged problem, water having to be taken at Swing Swang bridge as well as at the termini. Lastly, the smuts were said to contaminate the sand - certainly sparks set fire to a house at Leedon, despite spark arrestors.

It was not long before the Company sought alternative motive power and tried an Army surplus 40 HP petrol locomotives. This was successful and more followed. The steam locomotives were then redundant and after less than two years work sold in about 1922 to Bryant & Langford Quarries Ltd., Portishead, Somerset, where they ended their days. Henceforth all traffic was handled by internal combustion locomotives.

The third locomotive was obtained from the contractor. It was a 2½ ton 20 HP 4 wheel "Simplex" petrol locomotive built by the Motor Rail & Tramcar Co.Ltd., Bedford, works number 1856. It had been ordered by Lamb & Phillips on August 1st, 1918, "for immediate delivery" (i.e. two or three days, from stock) to Billington Road. Lettered L & P on each side, it cost £585. Their first order for spares was on June 12th, 1920 so probably it was at first hired by the L.B.L.R. according to one source) before being purchased. However, the first time the L.B.L.R. ordered Simplex

41

LIGHT RAILWAY LOCOMOTIVES

spares was February 15th, 1921, and these could have been for 40HP locomotives. Also an early photograph c1922/1923 which has the appearance of showing all the company's locomotives does not include the former Lamb & Phillips one, so it would appear that it was only hired, probably while the steam locomotives troubles were being rectified. Possibly the success of this little locomotive encouraged the Company to try larger petrol locomotives. They had no cause to regret the experiment.

The mainstay of the motive power for the next thirty years were 40HP 6 ton petrol locomotives built in early 1917 by Motor Rail for the War Office. By this stage of the War forward areas were supplied by a network of 60cm. gauge railways from standard gauge railheads. Mainline haulage on the War Department Light Railways was by steam locomotives, but their exhaust made them conspicuous, so close to the front the trains were handed over to petrol-engined locomotives. Motor Rail built about 260 of these 40 HP locomotives. Lighter work was undertaken by 20 HP petrol locomotives - about 600 being built by Motor Rail.

A 4 cylinder Dorman petrol engine drove a Dixon-Abbott patent 3 speed gearbox through an inverted cone clutch. The final drive was by chains to the axles. Speeds were 3.4, 8.2 and 10 mph. The driver sat sideways, over the centrally mounted engine, so that he could see in either direction. There were three types, open, weather-protected, and armoured, all based on the same four wheel frame whose side members, of deep steel plates, almost reached the ground. The sides were parallel only for the centre portion; the ends bent inwards. Heavy curved steel shields were fitted over the ends to protect the radiator and 20 gallon petrol tank and to provide weight. Additional ballast was provided by iron weights hung from each end of the frame. In the open type the driver was sheltered by an iron roof on four pillars. The protected type had a pair of double steel doors on either side between the end shields, and roof was fitted

LIGHT RAILWAY LOCOMOTIVES

with shields to help keep out the rain. The armoured type was like a little tank, having a flat armoured roof with small sliding vizors in the sides and ends of the "turret" for the driver to look out. Both protected and armoured locomotives had an additional ½" steel plate on one pair of diagonally opposite sides to cover ventilation holes in the bonnet, and the armoured ones had a similar plate over the roof. They were 11'0½" long over buffers, 6'6" wide over frames, 7'10" high, with 18" wheels on 4' wheelbase. The L.B.L.R. had examples of both the protected and armoured designs.

There are no L.B.L.R. locomotive records, so the following description of their stock is based mainly on photographs and the observations of a few enthusiasts since 1945. As most of the early locomotives appear to have been obtained through Army surplus dealers, Motor Rail's excellent records have not been able to shed much light. Only three were purchased direct from Motor Rail - one armoured type, later L.B.L.R. 1, on May 3rd, 1924. (Numbers do not appear to have been carried at first.) This might suggest that the first locomotives to replace the steam ones were numbers 3 and 4. A photograph whose probable date is about 1922 shows two engine sheds and the tracks to the third. These together could hold 5 40HP and 2 20HP locomotives. After the success of the first big petrol locomotive, the Company bought four more so it would seem that this was in fact its stud. The photograph mentioned shows two protected and one armoured 40HP, and two ex-W.D.L.R. 20HP locomotives.

About 1923 a L.B.L.R. 20HP locomotive which had been in an accident and bowed its frame, was sold to Joseph Arnold & Sons for use in their quarries. This might explain the purchase, from Motor Rail, of a reconditioned 20HP W.D.L.R. locomotive on March 12th, 1924.

The third locomotive purchased from Motor Rail was another 40HP one, bought on May 31st, **1934** for £260.

LIGHT RAILWAY LOCOMOTIVES

It had been reconditioned by them, having worked previously for Manchester Waterworks. This locomotive (number 5) presumably bought to help with the increasing traffic, was also presumably the cause of the shed nearest the main line being extended by 11'.

A.G.Wells visited the line in November, 1945, when there were six 40HP locomotives and two 20HP ones. A year later E.S.Tonks found six 40HP, four 20HP and two 40HP ones bought from Mowlem, the contractors, for spares. In May 1958 D.Semmens noted six 40HP and two 20HP (plus the modern locomotives) so that it would appear that the normal postwar stock, and hence probably the prewar stock too, was six large and two small locomotives. The two extra small ones observed by Mr.Tonks, could have been purchased to help with the heavy postwar traffic until more satisfactory locomotives could be obtained. Mr. Semmens discovered Numbers 4 and 6 (old) were the Mowlem's locomotives, number 4 then being used for spares and number 6 scrapped a year before, while the new 6 was also withdrawn as it was prone to derailments. The stock is summarised, as far as it is possible to do so, in the table which follows.

The 40HP locomotives had been converted in the late 1930's to run on Tractor Vapourising Oil (paraffin) after starting on petrol, which cut the fuel costs considerably, although some power was lost too. This enabled the locomotives to remain in use for as long as they did. There were plans for fitting them with diesel engines after World War II, but this was not done, modern diesel locomotives being purchased instead. However, some of the old locomotives were in use to the end of 1958, so completing forty years of service. On the L.B.L.R. the armoured doors were usually kept open for the driver's comfort and safety, and a tarpaulin hung across the opening. After 1945 electric car headlights were fitted, previously railway type oil lamps had been used after dark, hung from a central lamp bracket.

The 20HP locomotives, used for shunting and light trains, had a canopy in later years with tarpaulins for

LIGHT RAILWAY LOCOMOTIVES

extra protection. The driver's mate sat on the bonnet. Numbers were sometimes painted on the bonnet sides and in the early days the words Leighton Buzzard appeared. At first the livery was grey, later it was changed to green. Most of the large locomotives carried numbers in their later years, and some retained their old W.D.L.R. plates to the end. Latterly they had distinctive white markings on the ends for identification at Billingham Road (see "Operation"). In 1948 the cost price of the locomotive stock was £3015.

The Company's first new locomotive since 1919 was a 2½ ton 20/28HP diesel dispatched from Motor Rail's works on November 28th, 1950. No. 9 cost £742 and was built to a rail-gauge of 600mm i.e. 1'11⅝", not 2'. (The construction of all 2½ ton locomotives will be described in the chapter "Quarry Locomotives"). Electric headlights were fitted. However, the original engine gave a great deal of trouble and in February, 1951, was replaced by a new one. Two more new locomotives followed, both 6 ton 32/42HP models for main line work. No. 10 left Motor Rail's works on May 1st 1951, and cost £1325 complete with cab and battery lighting. The second No. 11, was ordered in October 1953, and delivered on 14th January, 1954. The price had risen to £1454.

In 1955 or 1956 three similar locomotives, but only weighing 5 tons, were purchased secondhand from the dealer G.W.Bungey Ltd., Heston, Middlesex. They were numbered 12-14 and had been built originally for a gravel pit near Wrexham worked by Sir Alfred McAlpine.

These large locomotives have frames of heavy steel channel, well braced. The centrally mounted cold starting solid injection two cylinder Dorman 2DL diesel engine develops 32 BHP at 1000 rpm and 42 BHP at 1500 rpm. A large diameter single plate clutch transmits the power to a patent gearbox giving three speeds 2.9, 4.75, 7.56mph at 1000 rpm. The final drive to the wheels is by heavy roller chains. The driver sits sideways at one end and at the other is a central longitudinal radiator. Hinged sheet steel covers

LIGHT RAILWAY LOCOMOTIVES

protect the engine. Cabs are optional on Motor Rail narrow gauge locomotives but all had standard cabs fitted. One side is open (closed by a tarpaulin) while the other is closed but fitted with a half height emergency and inspection door. Sliding windows are fitted back and front, and the roof is slightly ridged. Electric headlights are fitted at both ends.

The 5 ton locomotives have a tubular handrail across the front. This is not necessary on the 6 ton version which has cast iron ballast weights 2' high and 3 to 5 inches thick bolted to either end of the frame. Both models have flat transverse weights under the ends of the frame for ballast and to prevent damage to the gears when the locomotive is derailed, by limiting the distance it can fall. Overall dimensions are 8'8" long over frame (9'11¼" over buffers) 4'11" wide and 6'2¼" high. The 18" diameter wheels give a wheelbase of 3'4¾". The modern locomotives were painted medium green with the upper half of the cab white.

When the L.B.L.R. ceased to provide motive power in December 1958, the new locomotives were shared between Arnold's and Garside's. The old W.D.L.R. ones were dumped at the southern end of Billington Road yard and were scrapped in February/March, 1959, by Smith of Watford. He had some difficulty with the armour plate !

The L.B.L.R. was provided facilities for Motor Rail to test prototypes which were often photographed on the line for catalogue illustrations. To help reduce the length of locomotives so that they can be used in mines which have very narrow shafts, Motor Rail developed a lever brake model for export to South Africa where this problem particularly arose. In September 1953, the first such locomotive was built. This brake was first fitted to a normal locomotive in April, 1954, and No.21500 was tested on the L.B.LR. It is shown on the cover of the Motor Rail brochure with a train on the gradient up to the old locomotive sheds at Billington Road. In September, 1955, the prototype of a new design

LIGHT RAILWAY LOCOMOTIVES

of 50HP locomotives, No. 11001, was on trial. A photograph of it on one of Arnold's trains near Leedon was used in the brochure. Production models had a slightly different frame from the first four built. A standard locomotive, No.8724, built in 1940, and owned by the Motor Rail subsidiary Diesel Locomotive Hirers Ltd. was experimentally fitted with a 3 cylinder Perkins 3/152 engine, and automatic transmission. It was on trial in December, 1961 and January, 1962.

    The railway was used for publicity of a rather different sort in November, 1967, when a television commercial for Mars Bars was shot in Arnold's Double Arches quarry.

L.B.L.R. LOCOMOTIVES

			Wks.No.	Date			W.D.L.R. No.
-	0-6-0WT		HC 1377	1918	(a)	(1)	3207
-	0-6-0WT		HC 1378	1918	(a)	(1)	3208
-	4wP	2½T	MR 1856	1918	(b)	(2)	
-	4wP	2½T	MR 1757	1918	(c)	(3)	2478
-	4wP	2½T	LR 2213	1924	(d)		2496
1	4wP	6T	MR 3674	1924	(e)	(4)	2316
2	4wP	6TP	MR 1383	1918	(f)	(4)	3104
3	4wP	6TA	LR 478	1918	(g)	(4)	2199
4	4wP	6TA	LR 468	1918	(h)	(4)	2189
5	4wP	6TP	LR 3348	1934	(i)	(4)	2228
6	4wP	6TP	MR 1299	1918	(j)	(4)	3020
4?	4wP	6TP	MR 3675		(k)	(Scr)	3005
6?	4wP	6TP	MR 1283	1918	(l)	(Scr) 3004 Oct. 1957)	
7	4wP	2½T	MR		(m)	(4)	
8	4wP	2½T	MR		(m)	(4)	
9	4wP	2½T	MR		(m)	(Scr 1950)	A/G No.
9	4wD	2½T	MR 9547	1950	New	(5)	41
10	4wP	2½T	MR		(m)	(4)	
10	4wD	6T	MR 10272	1951	New	(6)	10
11	4wD	6T	MR 10409	1954	New	(5)	43
12	4wD	5T	MR 7932	1941	(n)	(6)	12
13	4wD	5T	MR 7933	1941	(n)	(5)	44
14	4wD	5T	MR 7710	1939	(o)	(5)	42

LIGHT RAILWAY LOCOMOTIVES

HC   Hudswell Clarke   A   Armoured   Scr   Scrapped
MR   Motor Rail   P   Protected   T   Tons

(a) Built for W.D.L.R. but not delivered. Regauged to 2' before purchase.
(b) ex-Lamb & Phillips Ltd., contractors. Hired c.1919-1920.
(c) ex? Delivered to France, August, 1918.
(d) ex-MR, 12-3-24, rebuilt of 1775 of 1918. MR purchased from Wm. Jones 1-4-22.
(e) ex-MR, 3-5-24, rebuilt of 595 of 1918, purchased from French Disposal Board. Originally armoured but rebuilt with new body.
(f) ex? Built 17-12-18. On L.B.L.R. by c 1922.
(g) ex? Built 26-3-18) One of these purchased
(h) ex? Built 26-3-18)   by L.B.L.R. c 1922.
(i) ex MR 31-5-34, rebuilt of 507, purchased by MR from Manchester Waterworks.
(j) ex? Built 29-7-18. Purchased c.1922.
(k) ex Mowlem 1946 for spares, but possibly taken into stock. Mowlem number 27, rebuilt of 1284 of 1918, delivered to France.
(l) as (k) but number 26, cabless. Delivered to Belgian Government.
(m) ex? One of these probably 9 was MR.849 of 1918. W.D.L.R. 2570.
(n) New to Sir Alfred McAlpine & Sons Ltd., Pant Farm Gravel Pit, Gresford, Wrexham, October 1941. On L.B.L.R. by 22-10-56, via Bungay.
(o) New to McAlpine, Pant Farm Gravel Pit, 9-6-39. Then to Derbyshire Stone Quarries Ltd., Hopton Quarry, then Hartington Quarry, then L.B.L.R. via Bungay.

(1) To Bryant & Langford Quarries Ltd., Portishead, Somerset c.1922.
(2) ? Returned to Lamb & Phillips c. July 1920.
(3) To J. Arnold & Sons c. 1923.
(4) Scrapped by Smith, Watford, c.February, 1959.
(5) To J. Arnold & Sons Ltd. 3-12-58.
(6) To George Garside 3-12-58.

QUARRY SYSTEMS CONNECTED TO L.B.L.R.

QUARRY SYSTEMS CONNECTED TO L.B.L.R.

Billington Road Depot & Pratt's Pit (Arnold)

928240   gantry
930242   pit
927242   washers

    Joseph Arnold & Sons Ltd. had two premises west of Billington Road. The southern-most, in use until 1969. was an elevated wood and steel platform on brick piers from which sand could be tipped into main line wagons. It was approached by a double track extension from the L.B.L.R. Horses were used to haul wagons up on to the "bridge" until 1938/1939. Later, shunting was the job of a 20HP locomotive. Wagons were pulled on to the tip and pushed back into the empties road. The usual tipping point in recent years has been on the inside of the bend of the gantry, the locomotive setting back to tip. The last method of operation was for the main line train to be divided and its locomotive gave banking assistance up the 1 in 15 gradient on to the tip for each half in turn. Since the closure of all the loops gravity shunting had to be done in the yard to run the locomotive round its train.

    Pratt's Pitt across the road was served by rail until August 5th, 1968, when a concrete road to the face was brought into use. In 1967/1968 the face nearest the L.B.L.R. was worked, and brought so close to the road that one loop had to be lifted and another closed. Sand was brought up either to the main line tip, or to a lorry tip beside Arnold's smithy (this required two reversals) or to a factory beside the main line tip. This factory was first a glass bottle works with sand brought in by L.B.L.R. then in the 1930's Grovebury Brickworks Co. used the site. This works took about 90 wagons a day from Pratt's Pit. About 1950 it was sold to a firm making jute sacks and the site was again for sale in 1961, although long disused tracks could still be traced in concrete roads and through thick grass. A ground level line was provided by the tip for incoming coal and stores, but it has not been used for years. Unwashed and bagged sand was handled at this tip (closed April 1969).

QUARRY SYSTEMS CONNECTED TO L.B.L.R.

Arnold's locomotive repair shops used to be housed in a corrugated iron shed approached from the L.B.L.R. terminal loops, but in 1958 the Company took over the former L.B.L.R. shops and the old building has since been demolished. The smithy and wagon repair shop are in a brick building beside the level crossing.

The washer stood north of Garside's depot in an awkward corner of Gregory Harris's old sand pit, adjacent to Garside's old pit. The approach from the L.B.L.R. was down a short stretch of about 1 in 25 gradient, across the road and round a very sharp checkrailed curve, which brought the track parallel to the road. Here the sand was tipped into hoppers. Loaded trains had a locomotive at each end, the main line one at the back ready to take the empties away, and a 20HP one at the front. This was attached at Page's Park loop where the large locomotive had run round. The 20HP locomotive stayed to bank the empties up to Page's Park and to catch any breakaways. Beyond the hoppers the line fell at about 1 in 10 to a reversing neck which gave access to the bottom of the washers. On this level was a shed for one locomotive, a tipping dock for loading main line wagons and a line to a dump for reject sand. When a horse was used for shunting below the washers its back was covered by a waterproof sheet. Occasionally washed sand was brought up the incline, two wagon loads at a time. The washers and screens were horizontal rotating drums of wire mesh. Much labour was involved and the plant became uneconomic, so in 1964 a new washer was built at Double Arches. This site was then abandoned and dismantled.

Billington Road Depot (Garside) 928241

The approach to George Garside's depot branched off the L.B.L.R. just before the loops and turned into a checkrailed curve to cross the road. Here the track ran through a brick shed, used to house the locomotives after work, and this gave access to a long level stretch. Loaded trains, with a 20HP locomotive leading and a 40HP one at the back were brought to a stand with the big locomotive just across the road. The 20HP one then shunted the train. The line led across a weighbridge and on to a wood and steel gantry from which main line wagons were loaded. Nearby were nine brick storage bins.

QUARRY SYSTEMS CONNECTED TO L.B.L.R.

A junction near the weighbridge led back towards the road. A siding led into one locomotive repair shop (later used by the Iron Horse) between the "shed" and an inclined ramp for loading lorries (rarely used). Another siding served a second locomotive repair shop. The main track passed two hoppers for the washers before entering the drying, grading and bagging plant built soon after the opening of the L.B.L.R. A trailing connection was made with the line serving the brick wagon repairs shop and smithy, outside which stood a wooden lifting tripod. At the far end of the bagging plant the track finished, very close to Arnold's line. Often one or two locomotives were stored there. The low level of the washer was approached by a long gently falling line which ran back from near the tipping gantry. It was laid with light rail. The line ended in a tip for rejects. By the washer two tracks diverged into some bushes near the edge of the old flooded pit. This was Garside's graveyard. From about 1950 to the early 1960's withdrawn locomotives were dumped here and left, often for years, before their remains, shorn of all useful components, were cut up. Ten or more locomotives, mainly un-identifiable, might be found here at one time.

With increasing quantities of sand being loaded at the quarries Garside's used the L.B.L.R. only for the sand which had to be dried. This was mainly dispatched by lorry too so there was little justification for any plant at Billington Road. Accordingly new premises were constructed at Double Arches, and Billington Road closed at the end of 1964. The Iron Horse Preservation Society were given the use of one shed and the rest abandoned. The level crossing and its approach were lifted but most of the track remained in situ until 1968 when the site was sold for re-development.

Chamberlain's Barn Quarry (Arnold)   926264

Joseph Arnold & Sons Ltd. have a large quarry - Chamberlain's Barn - on the northern edge of Leighton Buzzard beside Heath Road, opened c.1910. Much sand

QUARRY SYSTEMS CONNECTED TO L.B.L.R.

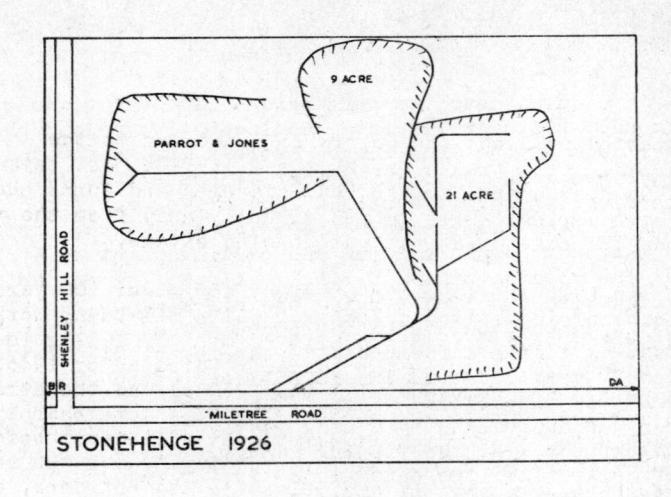

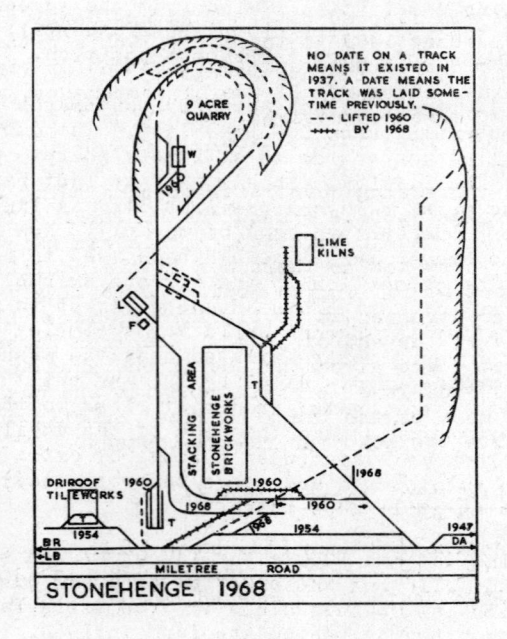

STONEHENGE 1926

STONEHENGE 1968

QUARRY SYSTEMS CONNECTED TO L.B.L.R.

leaves by road but the quarry is connected to the
L.B.L.R. The branch leaves the main line just beyond
the Co-op loop at Vandyke Road crossing and turns sharply
through a right angle to run directly away from the road
along the edge of a post war housing estate.

In 1926 the line ran straight for about 600 yards
before curving left into the pit which was being worked
southwards, close to Broomhills Farm. By 1937 a loop
had been laid close to the junction (disused by 1946, but
the points not removed until the 1950's) and the straight
run had been cut to 250 yards. The line bore sharply left
to avoid the face which it skirted at a distance before
running into the old part of the quarry nearer the road.
Here there were screens and, close to the entrance, a loop
by a tipping dock. Tracks led from the screens to the
face, and to tips in the old parts of the pit. A locomotive shed stood near its present position.

By 1947 the face had advanced southwards so that the
present reversing neck had been laid. The curve here from
the main line connection is extremely sharp - about 12 yards
radius. The L.B.L.R. finished at the last rail joint
before the point - beyond was Arnold's. A line continued
north along the eastern edge of the pit. Now right
beside the hedge, this line had been used to remove overburden. The connection to the screens skirts the southern
edge of the pit, which has advanced almost to the boundary
hedge. After the Second World War a complex group of
sidings for wagon storage existed near the washer, a large
two road engine shed - 41'6" by 18' was built of concrete
blocks with a corrugated iron roof close to the existing
wooden shed which became the home of the mobile compressor
used for rock drilling (when patches of extra hard sand
stone are encountered and needs to be blasted). Drilling
equipment is still kept there.

During the mid 1960's the layout at the screens was
simplified following the erection of a new washer. The
face, which in 1960 was close to Broomhills Farm, was
extended northwards behind the farm in 1964.

QUARRY SYSTEMS CONNECTED TO L.B.L.R.

Locomotives were introduced c.1925. In 1968 the quarry had five locomotives allocated to it, three in use, one spare and one converted to an air compressor. Two trains of 4 wagons operated between face and washer. One would be loaded by an excavator starting with the wagon nearest the locomotive and then hauled to the washer. The driver then left his train for the other which was pushed back for loading. A shunter moved the loaded wagon past the hopper, tipped them then took the empties on to the through line to leave the hopper road clear for the next loaded train. The third locomotive moved loaded wagons from beneath the washer up to the lorry tip and stock pile by the entrance. Some shunting beneath the washer was done by hand. Track in the pit was laid on wooden sleepers at approximately 3'6" centres.

In L.B.L.R. days their locomotives worked as far as the sidings by the screens. Some sand was still worked to Billington Road in 1969, some goes out to Stonehenge and Double Arches, hauled by $2\frac{1}{2}$ ton locomotives, but much of the sand goes by lorry to the nearby Anchor Tile works. In mid 1969 a direct road was made from the washer to the tile works, and the washer altered to accommodate lorries.

New Trees Quarry (Arnold) 931275

A new quarry was opened on the west side of Shenley Hill in the summer of 1963. It has not been on continuous production. The branch continues northwards from the Chamberlain's Barn reversing point between the quarry and hedge. An overgrown loop marks the summit of a gentle climb. At the field corner the track passes through the hedge to run on the eastern side and a quarter of a mile straight stretch of falling gradient brings the line to a crossing over Shenley Hill Road. Just before the crossing is a trap siding. This crossing is only $\frac{5}{8}$ mile by road from the L.B.L.R. crossing over this road but nearly three times as far by rail.

Photo: Bedford Silica Sand Mines Ltd)
Hand excavation of sand and barrow run for overburden removal.(1934)

Photo: Collection A. Keef)
Opening day of the LBLR, at Billington Road. This is the only known photograph of the LBLR steam locomotives.

Maker's photograph of Hudswell Clarke 060 WT built for War Department Light Railways. (Collection R.N. Redman)

LBLR Sheds at Billington Road I - about 1922 with 40HP (armoured and protected) and 20HP locomotives (left hand one is LR 2478) (Collection A. Eggleton)

LBLR Sheds II - Nov. 1945, 2 protected WDLR locomotive LR3104 stands outside the original depot. (Photo: A.G.Wells)

LBLR Sheds III - About 1956. 1948 building with ex WDLR 40 and 20 HP locomotives and a modern 40 HP 6 ton locomotive. (Photo: F.Jux)

Arnold 43 collecting empties from Leighton Buzzard Concrete Co. Ltd siding April 1961. (Photo: author)

Notice beside LBLR at Marley Tiles Ltd works, in English, German, Jugoslav, Polish. (Photo: author)

No.4 braking No.42 down to Billington Road sidings. WD bogie wagon of dried bagged sand next to loco. Jan.1962 (Photo: P.R. Arnold)

Swing Swang Bridge over Clipstone Brook. Arnold 7 on train from Chamberlains Barn. April 1961. (Photo: author)

Train headed by rebuilt 40 HP locomotive (No. 1?) showing proximity to Vandyke Road. April 1954. (Photo: G.H. Starmer)

Arnold Nos. 13 and 19 passing on the double track east of Stonehenge. July 1954 (Photo: F.H. Eyles)

Armoured locomotive 3 waiting at Stonehenge Brickworks. Note identification stripe on bonnet. Quarry stables behind locomotive. April 1954. (Photo: G.H. Starmer)

The end of the LBLR, with a train coming off Garside's branch - Arnold's branch diverges by the third wagon. Note the signal. Jan. 1961. (Photo: author)

Garside's Billington Road depot. Unloading bagged sand from bogie wagon. Cabless locomotive No.14 Devon Loch. Jan 1960. (Photo: author)

Empties crossing Billington Road hauled by No.12 - note identification symbol. About 1960. (Photo: F. Jux)

Chamberlain's Barn Quarry with No. 6 at the face August 1968. (Photo: author)

No.21, the boat-framed locomotive converted to a self propelled air compressor for rock drills. August 1968. (Photo: author)

Arnold's Double Arches Quarry: Hand loading in April 1954. Petrol boat framed locomotive No. 23. (Photo: G.H. Starmer)

Arnold's "bridge", Double Arches. Garside's old tip in left background, and an armoured locomotive hauls empties on the main line.
April 1954 (Photo: G.H. Starmer).

Bogie "type D" wagon used for bagged sand.
Oct. 1959. (Photo: M. Swift).

The last boat framed petrol locomotive No.21 Festoon.
Aug. 1968   (Photo: author)

Garside's Double Arches Quarry No.37 "Gay Donald" (with curved cab roof) fitted with snowplough.   A second plough lies behind.   April 1969.
(Photo: C. Daniels)

Iron Horse inaugural special about to leave Page's Park, 3 March 1968 with IHRR No.1. (Photo: Leighton Buzzard Observer)

Iron Horse Preservation Society special headed by "Chaloner", with "Rail Taxi" on the main line, at Page's Park. December 1968. (Photo: P. Nicholson)

Leighton Buzzard Tiles (1934) Ltd locomotive on the gantry.
April 1961. (Photo: author)

Leighton Buzzard Brick Co.Ltd. - Ledburn Road pits
about 1930. Note harness and the screen for sand
beside the wagon. (Photo: F.C. Rickard)

Leighton Buzzard Brick Co.Ltd. - Potsgrove pits. Haig out of use beside a pile
of lifted track. About 1950. (Photo: G. Alliez)

Bedford Silica Sand Mines Ltd tipping dock beside the main road in 1928.
Works in right background.   Note screw brake column on end waggon.
(Photo: Bedford Silica Sand Mines Ltd)

Woburn Abbey Parkland Railway.   No.1 in sylvan surroundings. Sept. 1968.
(Photo: F. Jux)

QUARRY SYSTEMS CONNECTED TO L.B.L.R.

Beyond the crossing is an elevated fuel tank and a loop, still on a falling gradient. A trap siding guards the steep descent into the pit, where the face is being worked eastwards.

All sand is taken to Double Arches for washing. As all the quarry extension is laid in light rail only 20HP 3½ ton locomotives are used, usually 34 and 41, one at each end of the train of about 16 wagons. This solves the problem of reversal at the Co-op loop. When returning with empties the rear locomotive is not coupled but follows a few yards behind the last wagon. The locomotives are now shedded at Double Arches after a period at Chamberlain's Barn. When this was the case the two 20HP locomotives, one at each end, hauled the wagons as far as the Co-op loop. There a 40HP locomotive took over. This had arrived on a train from Double Arches. A second 40HP locomotive brought empties for New Trees from Billington Road, and returned there with the train of washed sand. Thus the main line was worked in two halves, with both 40HP locomotives returning to Billington Road in the evening. However, when the fall in traffic required the attention of only one 40HP locomotive, the present system was instituted.

<u>Stonehenge Brickworks</u> 941274 <u>Twenty One Acres Quarry</u> (Arnold) 942276 and <u>Nine Acres (or Chance's) Quarry</u> (Arnold) 938276.

These quarries have been worked by Joseph Arnold & Sons for many years.

In 1926 there was a loop close to the L.B.L.R. and the quarry branch then divided in two. One line almost doubled back on itself and led to Parrot and Jones Quarry. The other turned through a right angle and served Arnold's Twenty One Acre quarry, opened early 1920, the face in use being near the eastern boundary of the site close to the later lime kilns. A building which could have been a locomotive shed is shown, three locomotives being the normal allocation. The quarry entrance was opposite Mile

QUARRY SYSTEMS CONNECTED TO L.B.L.R.

Tree Farm. After working out the deposits of silver sand the quarry was later partially filled in by Luton R.D.C. who used it as a tip. In 1926 Nine Acres appears not to have been in use.

Stonehenge Brickworks was built about 1935, originally trading as Vandyke Sand-Lime Bricks Ltd. Bricks are made by mixing sand and quicklime and then adding water to slake the lime. The reaction binds the mass into a hard white brick. Lime used to be made on the premises from local chalk, then it was obtained from Totternhoe Lime & Stone Co.Ltd. Ltd., and it is now received from Buxton in Derbyshire.

The brickworks are built at right angles to the L.B.L.R. The hoppers for sand and lime are in the northern side. Along the south-western is a stacking yard spanned by an overhead travelling crane. There is a quantity of 2'6" gauge track in this yard for the flat trolleys 3'4" long by 5'7" wide used in the brick-making plant itself.

In 1937 the layout was much the same as 1960. Nine Acres was in use again, working a lower bed of sand or "compo". The earlier track petered out in old workings, of Twenty One Acre, disused from the early 1930's. A new line made a trailing connection with the loop, turned to run beside the stacking yard and continued into the pit. Beside the stacking yard was an elevated siding to a tipping dock. A new facing connection for sand from Double Arches was laid into the L.B.L.R. and the track passed the brickworks hoppers. Near the L.B.L.R. there was a diamond crossing over the route to the old workings. By the hoppers was a loop, and a double track branch to the lime kilns. Beyond the loop the line curved to join the other route to the pit by the wooden engine shed. In 1937 there was a loop by the junction with a building spanning a short length of both loop tracks. No trace of either remained in 1960. The quarry tracks were on two levels, curving in opposite directions round the pit.

QUARRY SYSTEMS CONNECTED TO L.B.L.R.

When the lowest level of sand had been reached on the western edge of the pit the face was worked eastwards by dragline. A screen was built in the pit. Now sand is loaded at a high level siding. It is taken down a steep gradient, locomotive leading, to the screen, then brought out for dispatch. Waste screenings are tipped in the pit from a conveyor - formerly rail trucks were used. The layout was simplified in the mid 1960's.

Some sand is sent out over the L.B.L.R. but most leaves by road. It is tipped on to stockpiles by the brickworks entrance where a front loading shovel loads it into lorries. Three loaded wagons is the maximum that can be pushed up out of the pit by a $2\frac{1}{2}$ ton Simplex. Three locomotives are usually allocated here.

Next door to the brickworks immediately west of the line to the lorry tip is the factory of Driroof Tiles Ltd., where tiles are made from sand and cement. A siding makes trailing connection with the L.B.L.R. and on this siding is a loop which passes just inside the factory to serve a tip for sand.

In 1968 the brickworks took 50 wagons loads daily from Nine Acre pit, while Double Arches supplied 47 wagon loads for the brickworks and 20 for the tileworks. Locomotive 19 shuttles continuously to and from Double Arches with this traffic, hauling about 8 wagons each time. Driroof empty wagons themselves as required, but Arnold's drivers tip brickworks wagons on arrival.

Parrot and Jones, Shenley Hill 938274

Parrot and Jones, engineers, of North Street, Leighton Buzzard owned several small sand pits. The largest was between Nine Acre Quarry and Shenley Hill Road. In 1926 this was connected to the L.B.L.R. by a 300 yard long branch from the Nine Acre loop, and there was another 200 yards in the pit. This firm are said to have owned a 10HP Planet locomotive. Certainly there was a sufficiently long haul to justify one.

QUARRY SYSTEMS CONNECTED TO L.B.L.R.

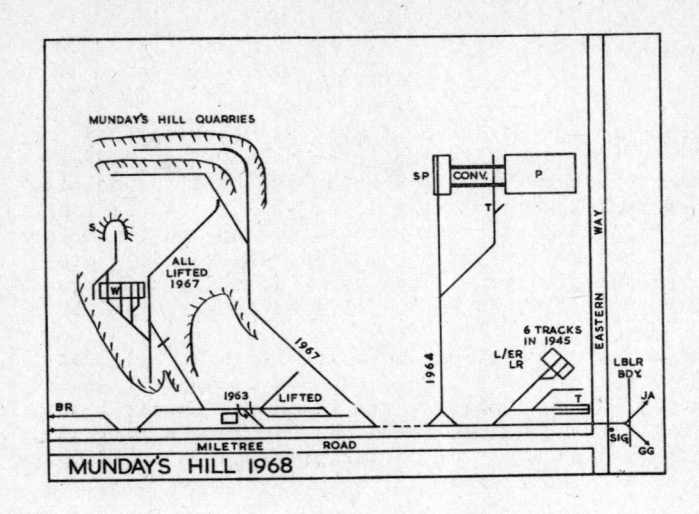

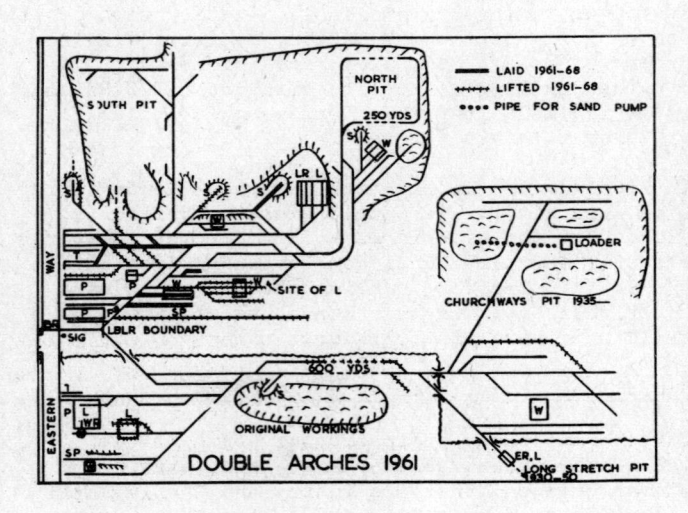

QUARRY SYSTEMS CONNECTED TO L.B.L.R.

By 1937 the pit was derelict, possibly a victim of the recession, and it is now very overgrown. Odd lengths of 20lb rail may still be found. Also in the undergrowth are a 2' gauge prefabricated point and an end tipping wagon chassis. One curiosity on the southern edge of the site close to a rough track is a length of 2'2" gauge track. This is probably a relic of a horse worked system before the L.B.L.R. was built.

Munday's Hill (Garside) 942279

George Garside opened this quarry c.1926. The early layout is not known but the site would suggest that it began as a pit close to the L.B.L.R. and gradually worked away into deeper cover. In 1960 a facing siding was thrown off the L.B.L.R. just beyond the end of the double track. This siding paralleled the main line and then divided into a loop and siding at a 3 into 1 point. By this point was another which made a trailing connection to the quarry line which descended at about 1 in 20. At the foot of the gradient, out of sight of the road, was a screening plant with several sidings. The main track continued to the face, one end of which yielded yellow sand and the other end white.

In 1963/1964 the layout by the L.B.L.R. was altered to a simple loop so that trains could run from the quarry, to the Double Arches washer without reversal. The screen in the pit was then closed and dismantled. A new connection was laid from the L.B.L.R. a few yards beyond the loop in 1966 and the old route abandoned, although the loop remains in use for locomotive storage while drivers rest in the brick hut by the road. The gradient out of the pit was reduced. Two faces are now worked at different levels, producing sand of a variety of colours from silver to red-brown.

This pit is worked in close conjunction with Double Arches which supplies locomotives. Operation will be described later.

QUARRY SYSTEMS CONNECTED TO L.B.L.R.

Eastern Way Installations (Garside)

943284 sheds
941284 drying plant

To replace the Billington Road plant, a modern drying and grading plant was built beside Eastern Way at Double Arches. It was brought into use at the end of 1964. A triangular junction, with both curved sides checkrailed, leads to a 200 yard line to the plant. The track soon divides, one to serve a tip and the other stockpile hoppers. Operation will be described later.

Just beside the level crossing at Eastern way is an elevated steel tipping dock and a two road corrugated iron shed. In 1945 it was a six road loco.shed. The tip was used for loading sand from Munday's Hill, but it has not been used since about 1945, lorries being loaded from gantries in Double Arches. The shed used to be Garside's navvy repairs shop. Since the move from Billington Road in 1964 locomotive repairs have been carried out here too. The two fitters keep a locomotive here which they use to reach any breakdown. Their locootive was an old petrol one, with a vice and toolboxes bolted to the frame to form a mobile workshop. However, as old equipment was replaced, breakdowns became less frequent and the vice was removed in mid 1967. A standard diesel locomotive, No.34"Kilmore", is now used. Remains of with-drawn lccomotives are now stored on the embankment to the tip.

Double Arches Quarries (Garside)

944285 depot
941292 pit washer

At the end of the L.B.L.R. the track divides. Garside's quarries are reached by the right hand branch. This crosses a small stream which marks the boundary of Arnold's and Garside's property, passes through a hedge and enters the

QUARRY SYSTEMS CONNECTED TO L.B.L.R.

quarry yard. The original pit, now flooded and used as a spoil tip, is at the further end of the yard. The 1926 map shows only a single track into the quarry.

In the early 1930's a new quarry – Long Stretch – was opened beside the Watling Street, and this was worked until 1950. Then c.1935 the present quarry – Churchways – was started, the workings being adjacent to Arnold's.

To reach the new quarries a railway was laid beside the old workings. For about quarter of a mile from the end of the yard the line is in a cutting through a wood. Then it curves slightly to run in the open to the junction for the quarries. In 1960 most of this main line, about 600 yards in all, was double track, but following the reduction of traffic in the late 1960's the empties road became relegated for use as a siding for surplus wagons.

Near the end of the double track the line divides. The right hand branch crosses a stream and leads to sidings serving the output side of the washer. One track continues, across a stream, and formerly served Long Stretch quarry. Following the closure of this pit a wooden shed was built over the track, just beyond the stream, for excavator spares, and from 1967 it was used to hold the last petrol locomotive, "Festoon".

The left hand branch also crosses a stream and divides again. The left hand line descends to the quarry and the right hand line branches into sidings to serve the input side of the washer. This washer was enlarged in 1964 to deal with Munday's Hill sand too. In 1960 sand was pumped to a hopper in the pit which loaded wagons but now an excavator is used.

The depot by the road contained a drying plant but since the opening of the Eastern Way drier and the move from Billington Road, the buildings contain

QUARRY SYSTEMS CONNECTED TO L.B.L.R.

only the wagon repair shop and the locomotive shed (formerly a drying shed), which supplies all Garside's lines in this area. The previous locomotive shed was a low two road corrugated iron erection, now gone. In addition, there are two elevated tracks, one serving a stockpile and the other a gantry by the gate for loading lorries. (The layout has not been greatly altered since the change.) A simple corrugated iron shelter forms a windbreak by the tipping position here and a similar one is to be found at the quarry washer.

Neglecting "Festoon" stored in the quarry, and "Kilmore" in the fitters shop, Double Arches had an allocation of 7 locomotives in 1968, of which about 5 were in daily use as follows; one from Churchways quarry to the washer; one shunted at the washer; two worked from Munday's Hill to the washer or to Eastern Way and one took sand from the washer to Eastern Way or to the lorry tip. Munday's Hill is now the busier pit.

The excavator at Churchways is operated by the locomotive driver whose train usually comprises 10 wagons. The track being level, it is safe for him to leave the engine brake off. The wagons are moved along for filling by pushing them with the excavator bucket which can be racked out to give a greater reach, as well as slewed. Trains from Munday's Hill are 6 or 7 wagons. Those to the drier from Churchways are 15 wagons, although 10 wagon trains were operated before the expansion of Munday's Hill. The lorry gantry holds 5 wagons.

This busy system is unlikely to be replaced by conveyor belts in the foreseeable future. The sand is worked selectively and conveyors to each feeding point would be expensive. Locomotives are cheap to run and have a long life.

QUARRY SYSTEMS CONNECTED TO L.B.L.R.

Double Arches (Arnold) 943285

Arnold's main quarries are served by the left hand track from the junction at the end of the L.B.L.R. By 1926 there were several loops in the quarry yard and a long line beside the road presumably for loading carts. The first workings were well away from the road, beyond the present engine shed. In 1930 the face was 600 yards long and 20' to 30' high, making it one of the largest sand pits in the country. South pit, near the road, had been opened before 1947 and is now the only pit, the north pit finished about 1960.

The main line cuts diagonally across the yard before resuming its former direction in a large marshalling yard. Trailing connections serve the drying and bagging plant.

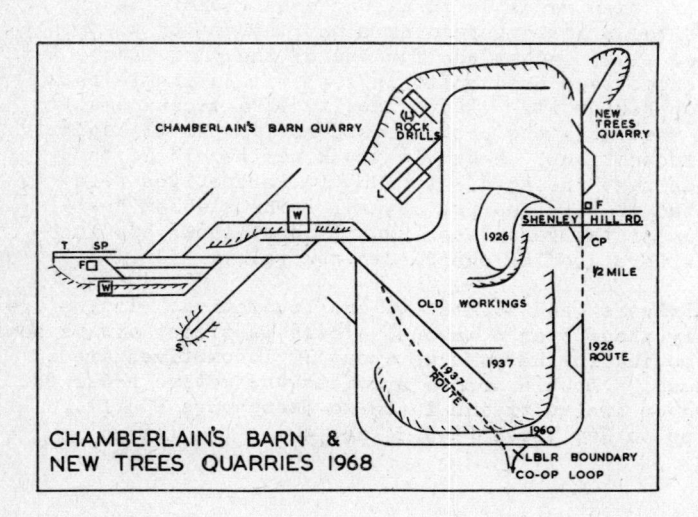

CHAMBERLAIN'S BARN & NEW TREES QUARRIES 1968

63

QUARRY SYSTEMS CONNECTED TO L.B.L.R.

The kilns used to be coke fired and fuel for them and coal for the excavators (introduced c.1930) was brought up in the bogie open wagons to a stockpile by the engine shed. The loading bays have verandahs which gives them a vaguely Continental appearance. Facing connections give access to the ramp for the new washer, erected in 1963 to deal with the extra sand from New Trees as well as to replace the old Billington Road washers, and to a siding for the washer sand dump, wagons usually being loaded by a tractor shovel instead of from the sand hopper.

Beyond the kilns, beside the road, are three lorry loading docks and bins for stockpiles. North from these, on the west side of the yard, is the line into the quarry which begins in cutting through an old spoil tip then descends in a long embankment. Beyond this is the primary screening plant, served by a steeply inclined line, and at the far end of the yard is the two road locomotives shed built in 1943 with a single road workshop beside it. There used to be a second small wooden shed near the present washer. The total capacity is 11 locomotives. A double track to the old North pit descends past the shed. In 1965 14 locomotives were allocated to this quarry, several being stabled overnight under the verandahs. (Until about 1950, 5/6 locomotives were used on overburden removal.)

Drivers tend to keep to one locomotive. Trains are normally about 5 or 6 wagons. There is almost always something moving in the yard as about 10 locomotives are in daily use. Double Arches provides the motive power for the ten or twelve trains daily to Stonehenge (No.19), and also to New Trees (Nos.34 and 40).

QUARRY LOCOMOTIVES

QUARRY LOCOMOTIVES

Horses were used for hauling wagons in the quarries, and in fact they lasted until about 1939 for taking wagons up on to Billington Road gantry. However, it was not long before locomotives were introduced. The first appeared in one of George Garside's quarries in November 1921. Four came in 1925, then three more in 1926, and by 1930 he had about a dozen at work. Joseph Arnold and Sons followed, obtaining their first locomotive secondhand from the L.B.L.R. around 1923. The progressive replacement of horses in the sand trade resulted in the closure of three saddlers in Leighton Buzzard.

The first locomotives, as well as many later additions, and the small ones on the L.B.L.R. were $2\frac{1}{2}$ tons 20HP petrol engined machines built by the Motor Rail and Tramcar Co., Ltd., Bedford. They were originally used on the 60cm gauge War Department Light Railways which served the Western Front during World War I. The design proved ideal for use on lightly laid lines in forward areas, and about 700 were built from early 1916. The works regularly produced 20-25 per week. After this war many of these locomotives were sold and passed into the hands of contractors, quarry owners, dealers etc.

These locomotives had a distinctive bowed channel under-frame in the shape of an elongated octagon. The frame had four stretchers, two at the ends which also supported small ballast weights, and two near the centre which carried the engine and gearbox. The axleboxes were carried on additional longitudinal girders between the end and central stretchers, thus leaving the centre section clear. A Dorman 2-cylinder 20HP petrol engine was mounted transversely on the left hand side of the locomotive, and drove via an inverted cone clutch, a Dixon Abbot patent 3-speed gearbox mounted on the right hand side. This gave speeds of 3.4, 8.2 and 10mph in both directions. The final drive was by roller chains to both axles. A distinctive sheet metal bonnet in two parts, hinged on the top, covered the engine, gearbox and the 14 gallon cylindrical petrol tank mounted above the latter. The W.D.L.R. number plate, aluminium

QUARRY LOCOMOTIVES

with separate figures rivetted on, was mounted on the bonnet sides. The driver sat sideways (so he could see both ways with equal ease) at the rear. His simple slatted wooden seat covered a toolbox, Within easy reach were the gear levers, a foot pedal for the clutch and a screw handbrake with a klaxon mounted in the pillar. No cab was fitted. At the front was a sideways mounted patent radiator not quite on the longtidudinal axis of the locomotive. On one side was the silencer and the other an empty space, so that a second man could ride on the locomotive. Four sandboxes supplied sand to each wheel. Central buffer couplings with a sprung buffer bar beneath them completed the rugged locomotive. Its basic measurements were 8'3" long over buffers, 4'10" wide (centre) 4' ends, 4'4½" high, 1'5¾" wheel diameter with 3'6½" wheelbase. This design continued to be built by Motor Rail until about 1926. While various improvements and modifications have been made, the basic layout and dimensions have altered very little in subsequent designs of 20-50HP locomotives. Straight channel sideframes were introduced in the mid 1920's, but replaced about ten years later by deep plate side frames. A heavy casting with several slots for different coupling heights, large diameter single plate clutches and a pressed steel driver's seat became standard in the Twenties. Two speed gearboxes giving speeds of about 3½ and 7 mph became the rule.

Diesel engines were introduced around 1935, but petrol-engined locomotives continued to be built for some time. The 2-cylinder Dorman diesel engine with cylinders 115mm bore by 130mm stroke (developing 20HP at 1000rpm or 28HP at 1600 rpm) had about half the fuel consumption of the equivalent petrol engine. Many petrol engined locomotives were converted to diesel by the quarry companies.

Cabs were (and are) optional on small Motor Rail locomotives. The standard cab had one side open, one closed by a door for maintenance and emergencies. There were narrow horizontal windows back and front and a ridged roof. Garside's locomotives usually have the

QUARRY LOCOMOTIVES

standard Motor Rail cab if anything, but most of Arnold's locomotives have very similar, but home-made cabs. Two of Garside's secondhand purchases had cabs with sem-circular roofs. Awnings – a simple roof on light supports – were once common. Tarpaulin sheets might be hung from the roof to give extra protection. A square hole was often cut in the sheet to act as a window.

Flat cast iron weights could be bolted under the ends of the frame to increase the locomotive weight from the basic $2\frac{1}{2}$ tons. This was to increase the maximum load that could be hauled, and the ballasted locomotives tend to be those working over extra steep gradients. Handrails are fitted across the front, and often to the radiator top, for a second man to ride on the locomotive.

Whistles were fitted to some locomotives. Petrol-engined ones had the whistle fitted to the cylinder head and worked off the compression. On diesels the whistle was mounted on a hollow box with a hole in one face. The box was pivoted so that the hole could be brought over the end of the exhaust pipe, thus causing the waste gases to pass through the whistle.

Although Motor Rail had abandoned bow frames from about 1926, such locomotives could be obtained from another manufacturer for another 15 years. Kent Construction & Engineering Co.Ltd., Ashford, Kent, bought a large number of ex-W.D.L.R. Simplex locomotives and spares soon after the First World War. The locomotives were reconditioned and sold as "reconditioned Simplex locomotives". Some of the spares were used to build "Planet" locomotives (designed by Kent Construction) but based very closely on the Simplex. Kent Construction ceased to trade in 1926. Eventually their stock and drawings passed to F.C.Hibberd & Co. Ltd., formed about 1930, whose works from 1932 were at Park Royal, London. Hibberd's continued to sell reconditioned Simplexes and Planets. Some were advertised as Planet-Simplex locomotives, an infringement of Motor Rail's

QUARRY LOCOMOTIVES

trade name, until a legal action prevented them from so doing. Hibberd's "Simplex" locomotives all had bow frames and were virtually undistinguishable from the Motor Rail "Simplex". The length, maximum width and height were all increased slightly to 9', 5' and 4'6" respectively. New radiators and sandbox lids were fitted as Motor Rail ones had "Simplex" cast on them. Other makes of engine, e.g. National, were often used instead of Dorman, standardised by Motor Rail. Where the two types of locomotives have worked side by side, a complete interchange of parts has often occurred. Four Hibberd "Simplex" locomotives worked in Arnold's quarries, and with a second hand Jung (Germany) which worked for a time in Garside's quarries about 1955, formed the only deviations from the Motor Rail monopoly.

The first diesel locomotives in the quarries were obtained in the mid 1930's and proved far more economical than the petrol ones. Arnold's converted one petrol engine to TVO but it was not a success, and so the petrol engined fleet lasted until after the Second World War. Some had diesel engines fitted into the original frames, but more often the withdrawal of a petrol locomotive followed the introduction of a secondhand diesel one. In 1968 both companies had only one boat framed locomotive left, and these were hardly used.

Both companies have tended to buy secondhand locomotives in preference to new ones. There was a considerable influx, mainly of petrol engined ones, in the late 1940's and early 1950's. In the late Fifties and Sixties, diesel-engined replacements for petrol, and petrol conversions, were obtained.

Arnold's tended to scrap their withdrawn locomotives quite quickly, but Garside's dumped their withdrawals on a pair of sidings behind the Billington Road washer. From time to time locomotives here were broken up, and all useful spares were removed. But this dump of derelicts (27 in May 1958 and 14 in April 1961) existed throughout the 1950's and was not cleared until 1962. Both firms continue to keep the frames of two or three locomotives

QUARRY LOCOMOTIVES

spare, in case of accidents. Garside's took two frames up to Double Arches when they left Billington Road, and a new line of derelicts (8 in August 1968) is forming on the embankment to the old tipping dock beside the present workshops.

Both companies have bought locomotives secondhand for spares. Arnold's especially do this when a replacement crankshaft is needed. This practice results in a number of locomotives being seen which have not in fact worked in the quarries.

While most of the locomotives used were 2' gauge, some were 60cm (1'11$\frac{5}{8}$") gauge. They may have been re-gauged when new wheels were required, but in the meantime they probably ran without difficulty as manufacturers of light railway equipment tend to give their products wide wheel tyres to offset the variations of gauge often found on industrial and contractors railways.

The last of Arnold's petrol locomotives - No. 21 - was converted in 1962 to a diesel self propelled air compressor for rock drills, primarily for use in Chamberlain's Barn Quarry. The compressor is mounted on the right hand side, in the space normally occupied by the fuel tank which was moved to the rear. The air reservoir is mounted beside the radiator. The air plant appears to have been salvaged from a trailer compressor, formerly used in the quarries,whose 10HP engine was insufficiently powerful. Garside's last petrol locomotive, also No. 21, is now stored, after some years as a mobile workshop for the fitters when it had a vice and toolboxes on the frame.

When the L.B.L.R. ceased to provide motive power its modern locomotives were shared between Arnold and Garside. Arnold's took three large and one small locomotives, and Garside's two large. However, with the closure of Billington Road depot in 1964, Garside's no longer needed their main line locomotives which were therefore sold to Hopkins-England Ltd. of Woburn Sands in 1965. (This firm scraps Garside's engines). After overhaul by Motor Rail they were resold to a firm in Singapore in August 1966.

QUARRY LOCOMOTIVES

Arnold's tend to use their largest locomotive – 43 – for main line haulage in preference to the 5 ton 42 and 44.

All locomotives are well maintained and painted medium green with white round the cab windows. Numbers (and names) are painted in white on the bonnet sides, and Arnold's have the number in black above the cab windows too. It would appear that at first locomotives were numbered in the order of purchase, and any blanks existing through withdrawals were filled before creating new numbers. Most of Garside's locomotives are named after racehorses which have won important races. The names may be changed when the locomotive is shopped, the foreman asking Mr. Delafield for his suggestions. This is in addition to "bonnet swapping". Deliberate changes of names have been more common in recent years.

However, complications have arisen. With running numbers on the easily removed bonnet covers and cab, if two locomotives pass through the shop simultaneously, it is possible for them to exchange identities. Arnold's did not scrap all their old locomotives, as new (i.e. second-hand) ones were obtained, and so managed to build up since about 1951 a small reserve of serviceable locomotives kept at the workshops. Thus for some years when a quarry locomotive comes in for repair an unnumbered spare has maybe been given its number and sent out straight away. In which case, after the repairs have been completed. the old number is painted out, and the locomotive goes into reserve until it is sent out as a replacement for another locomotive whose number it then takes. Locomotives average about 4 years between major overhauls, so a complete renumbering can take place in the natural course of events. However, this does not seem to have happened to any great extent, yet !

Then possibly changes of identity are further confused by the lack of works plates on many locomotives, especially the older ones. The plate on Simplex locomotives is usually rivetted to the frame – the rear engine bearer – but early ones had it on the front stretcher by the radiator. The works number may usually

QUARRY LOCOMOTIVES

be found stamped also at the base of the brake column but this was not generally known when there were large numbers of plateless locomotives in the quarries, so they are largely unidentified. Even some of the existing locomotives have only recently been identified. An example of the problem is the case of an ex-W.D.L.R. locomotive in Garside's dump which had no plate and carried a different name and number each side, with a third name visible beneath the paint on one side ! Then there is the case of No. 34 "Kilmore" (MR 7105) - the Double Arches fitters' locomotive. This had the frame from No. 35 "Doutelle", the bonnets of No. 34, the cab off No. 13 and a reconditioned spare engine. The frame of the original "Kilmore" lay outside, dumped, sporting the bonnet (and name) of "Doutelle".

The locomotive list has been compiled from Motor Rail records and the observations of enthusiasts. It is probably as near the truth as can ever be obtained. The difficulties in compiling it can be attributed in part to the advantages of standardisation ! It will be noticed that allocations to the quarries are remarkably permanent.

Joseph Arnold & Sons Ltd.

No.	Frame.	Former Nos.	Weight if not 2½ tons.	Motor Rail Works No.	F/n	Petrol/ Diesel.	11/45	5/58	4/61	5/65	8/68	Location
1			3½	7188	a	D		DA		BR	DA	
2	b			1169	b	P	9A	CB			Scr/60	
2		17 until c/65		4707	c	DxP	CB	CB	CB		DA	
3				5881	d	P	DA	DA	DA	BR	Scr9/59	
3		Gertie				D					DA	
4						DxP	BR	BR			Scr/60	
4		3 until c/66	3½	7201	e	P		DA		DA	Scr/60	
5	b					P	DA		DA		BR	
6	b					P	CB		BR		Scr/63	
6				7403	f	D		CB	CB	CB	Scr/56	
7				3862	g	DxP	DA	BR	BR		CB	
7	b			8723	h	D				BR	Scr5/61	
8		Lou 11		3996	i	P		to Stone Pits Aylesbury.			BR	
9	b					P	DA	DA	BR		Scr/63	
10	b	Bertha		341	j	DxP	9A	9A	9A	9A	Scr/66	
11	b			2288	k	P	DA	BR	BR		Scr/63	
12	b		FH.	1757	l	P	DA				Scr/46	
12	b	1 Winnie in 1945				DxP	BR	CB			Scr/60	
12				8732	m	D					Scr spares.?	
13	b					P					Scr/60	
13				8683	n	D		DA	CB	DA	DA	
14				1704	o	P	DA	DA			Scr/60	
14		32	FH	2161	p	D			BR		Scr/61	

72

							Location			
No.	Frame.	Former Nos.	Weight if not 2¼ tons.	Motor Rail Works No.	F/n	Petrol/ Diesel	11/ 45	5/ 58	4/ 61	5/ 65	8/ 68
15		19 until/63		4303	q	DxP	BR	DA	DA	CB	CB
16				4709	r	DxP	DA	DA	DA	DA	DA
17				3994	s	D					CB
18_H				8720	t	D					DA
19		18 until/63		4305	q	DxP	DA	DA	DA		DA
20	b			999	u	P	BR	DA			Scrc/60
20				8748	v	D	DA	DA	DA	DA	DA
21	b				w	DxP	9A	DA	CB	DA	CB
22	b					P	DA	DA			Scrc/60
22				8727	x	D		DA	DA	DA	DA
23	b					P					Scrc/56
23				7128	y	D		DA	DA	DA	DA
24						P		DA	CB		Scrc/63
25						P		CB			Scrby/58
25				7214	z	D			CB	CB	BR
26	b	13,4	FH	1917	aa	P	DA	BR	DA	DA	Scrc/63
26		15 until/63		4701	bb	DxP	BR	CB	CB	DA	DA
		19 until/65									
27	b	Hell's Angels (until/65)				DxP		9A	9A	9A	Scr/67
27						D					9A
28						P		DA			Scrc/60
29						P		DA			Scrc/60
30			FH	1922	cc	D					Scrby/58

73

Joseph Arnold & Sons Ltd.

No.	Former No.	Wt.	Works No.	F/n	P/D	11/45	5/58	Location 4/61	5/65	8/68
30						DA				Scrc/60
30			8695	dd	P		DA	DA	DA	DA
31	This number has always been blank, reason unknown.									
32 b										Scrc/57
33	4 in/65		7037	ee	DxP		BR	BR	9A	9A
34_H					DxP		9A	9A		Scrc/63
34_H b										
34	41 until /65	3½	9415	ff	D			CB	CB	Scr/65
35			9547	gg	D		DA	CB	CB	DA(NT)
36			7126	hh	D		CB	DA	DA	DA
37			8756	ii	D		CB	BR	CB	CB
38	39 in /58		8540	jj	P		DA	DA	DA	Scrc/63
39	38 in /58				P		DA	DA	DA	Scrc/60
40_H			7153	kk	D		DA	DA	DA	DA
41_H		3½	5859	ll	D				BR	DA(NT)
42_H		5	7710	gg	D			BR	BR	BR
43_H		6	10409	gg	D			BR	BR	BR
44		5	7933	gg	D			BR	BR	BR
S		3½	7215	mm	D				BR	Scrc/66
S			8597	nn	D				BR	BR

74

									Location	
No.	Former No.	Works No.	F/n	P/D	11/45	5/58	4/61	5/65	8/68	
	2	8700	oo	D				DA	BR	
	28	8724	pp	D				BR	Scrc/66	
S		9418	qq	D				BR	BR *	
S		9409	rr	D				BR	Scrc/66	
S		20558	ss	D					BR *	
71	Total (minimum)				22	35	35	36	34	

Known to have worked here, but not identified in the above list due to missing works plates.

	916	tt	P
	4705	uu	P
	4708	vv	P
	5073	ww	P
	5863	xx	D

* Frame only   b   boat frame (known). Remainder, except for
H   Headlamp and whistle   5 & 6 ton locos. mainly straight plate
S   Probably bought for spares   frames, probably with boat frames on
                                          earlier locos.

Location

BR   Billington Road   CB   Chamberlain's Barn   DA   Double Arches
9A   Nine Acre   NT   New Trees

Note: Spare locomotives are stored at BR. In 1968 only two BR locomotives were in regular use, 4 and 43, and then only as required.

Notes for Joseph Arnold & Sons Ltd. Locomotives.

Note: PLH & DLH - Petrol/Diesel Loco Hirers, Motor Rail subsidiary companies.

(a) desp. 1/7/37 to Ham River Grit Co.Ltd., New Pit, Woodley, Reading, purchased from H.R.G. Bletchingley c 1/65.

(b) desp. 10/18 to France, W.D.L.R. 2890, by 2/20 Groupement des Houilleres Envalier.

(c) New for PLH stock 1936, but sold to J.A. 10/36.

(d) desp. P.L.H. 16/3/36 (built as diesel) to HRG 23/7/35 ex HRG 1/65 as HRG 108.

(e) for Sir Lindsey Parkinson, Royal Ordnance Factory, Chorley, later at Cementation Ltd., Doncaster, ex Kennel, Hayes, 8/59. There is a bad figure on the plate, but from M.R. records 7001 is unlikely.

(f) DLH. sold to Land Reclamation Ltd. 7/7/48 purchased c 1956.

(g) desp. 21/6/28 PLH Reconditioned 20 H.P., WDLR 4409 (M.R. 371) "Water tanks to be removed and sander gear fitted". Perhaps this locomotive had rail washers instead of sanding gear. Clean wet rails give nearly as much grip as clean dry ones. Ex Chittenden and Simmons Ltd. Allington.

(h) desp. to Sir Robt. McAlpine, Hayes, Middx. c 1960.

(i) ex MR 20/10/33 reconditioned 20HP loco. £152 bought on H.P.

(j) del. France 4/17 as WDLR 1742 then by 11/26 owned by McCreadie Corporance Madaggart, then J.Summerville.

(k) Built 1940. Ex Holloway Bros. (Contrs) Ltd., London.

(l) ex L.B.L.R. c 1923. Orig. W.D.L.R. 2478 desp. France 8/18.

(m) desp. 7/6/41 to Ministry of Supply, Forrestry Unit, Carrbridge. Sold to M.Maclean, Holt Road, Cromer 10/57, to J.A. early 1960's for spares ?

(n) desp. 14/11/41 to Slindon Gravel Co., Slindon Common, nr Arundel, to J.A. c 1958.

(o) to France 7/18. W.D.L.R. 2425.

(p) desp. 30/11/38 to Royal Naval Armaments Depot, Ernesettle, Plymouth. Paxman 25 HP 2 RQT engine, to J.A. c 1958.

(q) del PLH 3/12/34 to J.A. 11/10/37.

(r) NEW to J.A. 31/10/36 £200.

(s) desp. 3/4/46 to Sir Robt. McAlpine, Hayes (order of 10 at £560) received by MR 10/56, to JA c 1963.

(t) desp. 21/3/41 to Royal Ordnance Factory, Burghfield, Reading. Purchased from McAlpine c 1960.

(u) del. France 8/18 W.D.L.R. 2720. At Establissement Decauville, Aisne, Paris 2/26.

(v) desp. 17/2/42 to Sir Alfred McAlpine, Hawthorn, Wilts. Order of 12 at £478 with exhaust quenchers and washers. Purchased by J.A. from R. Ouse (Yorks) Catchment Board 1/60.

(w) rebuilt to self-propelled air compressor 1962.

(x) desp. 22/5/41 to R.Ouse (Yorks.) Catchment Board, Ricall, Selby. £458 to J.A. 1/60.

(y) DLH. Sold to Land Reclamation Ltd., Pentonville
     Road, London 9/6/38, 3½ ton when built, to J.A.
     c 1960.

(z) DLH. 28/2/38.

(aa) NEW to J.A. 1/5/35. National 2D engine.

(bb) NEW to J.A. 18/2/36.

cc) del. 11/5/35 to Holloway Bros., Millbank, London.
     Dorman 2 JO engine, to J.A. in 1950's.

(dd) As (v) but del. 26/12/41. order of 19. Purchased
     by J.A. from R.Ouse (Yorks) Catchment Board 1/60.
(ee) PLH. to J.A. ex McAlpine c 1960.

(ff) desp. 27/4/49 to Holloway Bros. (London) Ltd.,
     Millbank, 3½ tons when built, to J.A. c 1964.
     Scrapped mid 1965 following a collision.

(gg) ex L.B.L.R. 12/58. 43 for sale 7/69.

(hh) desp. 8/4/36 Richard Briggs & Sons, Bankfield Lime
     and Roadstone Works, Clitheroe, 3½ ton to J.A. 1950's.

(ii) desp. 24/2/42 to Sandstone Ltd., Kings Lynn, £482
     By 10/46 at Cromhall Quarries, Charfield, Glos.
     ex Bungey c 1955.

(jj) desp. 25/7/40 to War Office. Order of 54 60 cm
     gauge ex British Waterways, Southall 1958.

(kk) desp. 1/1/37 to R.Ouse (Yorks) Catchment Board £345
     to J.A. 1950's.

(ll) desp. P.L.H. 28/5/34 (as diesel with aluminium pistons)
     13/7/34 sold to HRG. ex HRG c 1/65 No 110.

(mm) desp. 9/3/38 HRG. Ham. 3½ ton ex HRG c1/65 No.11

(nn) desp 19/3/42 DLH . Sold 12/6/46 to BSS. Purchased
     JA. c 1966.

(oo)   as (v) but del. 31/12/41. Order of 19 ex
       R. Ouse (Yorks) Catchment Board 1/60.

(pp)   as (h) ex MR (hire loco) c 1964.

(qq)   desp. 14/6/49 to Holloway Bros. (London) Ltd.
       3½ tons to J.A. c 1964.

(rr)   del. 7/12/48 to Holloway Bros. (London) Ltd.
       3½ tons to J.A. c 1964.

(ss)   desp. 7/4/55 DLH. Reconditioned and sold 26/8/60
       to British Waterways, Brimsdown, to Ham and Hall
       Ltd., Woodside Brickworks, Croydon, to J.A.
       1967/68.

(tt)   del. 6/18 to France W.D.L.R. 2637.

(uu)   desp. 24/11/36 to J.C. Oliver Ltd., Airedale
       Plant and Rly. Works, Leeds.

(vv)   PLH, later converted to diesel.

(ww)   NEW 6/5/30.

(xx)   NEW 31/5/34. £330 H.P. £75 + 12 monthly payments
       of £21. 5. 0.

George Garside (Sand) Ltd. Locomotives

Location (and name if altered)

No.	Frame.	Name & previous No/name.	Motor Works No.	foot note.	Rail/Pet/Dies.	11/45	5/58	4/61	5/65	8/68
1	b	Benghazi	374	a	P	DA	br	br		Scr/61
1	b	(21?)			P		DA	br		Scr/61
2	b	Steady Aim	1044	b	P	BR	br			Scr/60
2		Nickel Coin			P		DA	br		Scr/61
3					P					Scr
3		Migoli			P					Scr
4		Airborne			P	G	br			Scrby/58
5		M/Love (Colorado Kid)	3795	c	DxP	DA	br	br		Scr/61
6		Brendan's Cottage			P	DA				Scr/61
						GM⎱	MA⎱			Scrby/58
7		M'sieur l'A miral	4019	d	P	DA⎰	br⎰			Scr/60
		(Hellenique, 8 Golden Miller)				H				
8		Golden Miller (Hellenique)			P	DA				Scrby/58
8					P					Scr
9		Supertelo (?)			P	G	br			Scr/60
10	b	Cider Apple	3828	e	DxP		br	br		Scr/61
10		Tearaway			D			G		Scr/63
10			10272	f	D			BR	BR	Sold/65
11		El Alamein	5002	g	D	BR				Scrby/58
12			7932	h	D		BR⎱	BR⎱		Sold/65
13		Retrial (Lemon Cheese)	5870	i	D	LC⎰	BA⎰	BA⎰		Scr/66
						AX⎱	FD⎱	FD⎱	D⎱ A
13*		Arkle (17 Damredub,	7152	j	D	DA⎰	G⎰	G⎰	G⎰ DA
		Alexander, French Design)								
14		Devon Loch	7492	k	D					da

80

No.	Frame.	Name & Previous nO/name.	Motor Works No.	Foot Note.	Petrol Diesel.	11/45	5/58	4/61	5/65	8/68
15*		Brown Jack (Hyperbole Tulyar, Much Obliged)	7148	1	D	BR	MO)DA)	MO)DA)	MO)DA)	BJ)DA) An
16		Anglo (Pasch, Three Cheers,Auriole,Ribot)	7149	1	D	P BR	R BR	R BR	R DA	G
17		Alexander			P		br			Scrc/60
17		Damredub	7036	m	DxP					G
18		Langton Abbot			P		br	BR		Scrc/60
18		Honey Light			D		BR	BR	br	da
19	b	Lovely Cottage	4568		P		BR			Scrc/63
19		(10 Foxtrot)		n	P					Scr/65
20					P					Scr
21					P		br**	DA**		Scrq,'50
21	b	Tosca (5)			P	BS) BR)	DA F) G)	F) G)	F)** F) DA)	Scrc4/65 F
21	b	Festoon (12Festoon, Black Speck)	4570	o	P	3) DA)				DA +
22					P		br	br		Scr
23	b	(3 Monty)			P		br	br		Scrc/61
24	b	Sheila's Cottage			P					Scrc/61
25					P					Scr
25	h	Alycidon			P		DA		br	Scrc/61
26		Scratch II	5011	p	P		br	br		Scrc/60
26					D		BR)DA)	DA 36)	DA 36R	Scrc/60 Scrc/67
27		(Torch Singer,Artic Prince)					36)DA)	36)DA)	DA)	27)
27ᶜ		(36 Relko until c/67)	5852	q	D					DA)

81

George Garside (Sand) Ltd. Locomotives.

Location (&name is altered)

No.	Frame.Name &Previous no/name.	Motor Rail Works No.	Foot note.	Petrol Diesel.	11/45	5/58	4/61	5/65	8/68
28w	Flush Royal	8917	r	D		G	G	G	DA
29	Ayala (Quare Times. Supreme Court)	7374	s	D		QT} G}	QT} G}	A} DA}	A} DA}
30	Fleeting Moment		t	D		DA	DA		Scrc/62
30w	Larkspur (10Larkspur Tearaway, Uncle Joe.)	7195	u	D	TJ} DA}	T} G}	10 L} G}	30L} G}	30L} G}
31	Team Spirit (Oxo, Good Taste)	7371	s	D		GT} DA}	0} DA}	TS} DA}	TS} G}
32	Hard Ridden (Gay Donald, Teal)	7372	s	D		GD} DA}	HR} DA}	HR} DA}	HR} da}
33*	Utrillo (Crepello,Royal Tan, Wilwyn)	7140	v	D		c} DA}	α} DA}	u} DA}	j} DA}
34	Kilmore (35 Doutelle)	7105	w	D		D} BR}	D} BR}	D} BR}	K} DA**
35	Doutelle	8713	x	D					da
36	Relko (14 Devon Loch Halloween, Sun Chariot	7145	y	D	SC} DA}	DL} BR}	DL} BR}	R} BR}	R} da}
37	Gay Donald (Hard Ridden)						GD} DA}	GD} DA}	da

82

No.	Frame.Name &Previous no/name.	Motor Rail Works No.	Foot note	Petrol Diesel	11/45	5/58 S	4/61 D	5/65 K	8/68 GD
37c	Gay Donald (34 Kilmore Sundew,Darius)	7108	z	D		DA	DA	DA	DA
50	7	3850 Jung	aa	P			br		Scr/61
69		5215	bb	D					Sold
57 total									

Locomotives known to have worked for George Garside, but not identified in list due to missing plates

		1107	cc	P					Scr
		3789	dd	P		br			Scro/60
		5008	ee	P					Scr
b	3 frames, unidentifiable			P			br		Scro/61

Locomotive bought for spares

		4808	ff	D		br	br		Scro/65
		4809	ff	D		br	br		Scro/65
		5864	gg	D		br	br	br	da
		7115	hh	D					da
		7414	ii	D					da
6		8725	jj	D		br	br	br	da
	totals			W/O	17	24	24	19	13
				Der.	–	18	17	3	10
63	Total (minimum)				17	42	41	22	23

Notes:

* * Have frames out of true following collisions so have extra packing in springs.
* ** Fitters' locomotive. "Tosca" and Festoon" formerly fitted with vice and toolboxes to form mobile workshops.
* + in store
* c curved cab roof
* w withdrawn early 1969 after transfer to DA.
* b Known boat framed loco. Remainder, except for 10 & 12, mainly straight plate frames and probably boat frames on early locomotives.

Location

BR Billington Road. DA. Double Arches including Munday's Hill. G Grovebury. br, da denote locomotive withdrawn or derelict at this location. The other letters in these columns are the initials of the name, if it has been altered. Stock at Grovebury in 8/68 was moved to DA in spring of 1969.

Notes for George Garside (Sand) Ltd. Locomotives

(a) del. 5/17 to France W.D.L.R. 1775, to GG from M.R. 15/11/21.

(b) del. 8/18 to France W.D.L.R. 2765 (see aa).

(c) del. 12-7-26, rebuild of M.R. 219, W.D.L.R. 219, 60cm to GG 6/26.

(d) del. 25-1-26 to PLH, sold to M.R. 28-4-28 later to GG (/28?)

(e) rebuild of M.R. 1821 W.D.L.R. 2542, to GG from M.R. 20-3-26, £365, 60cm. Converted to diesel 9-12-35.

(f) ex L.B.L.R. 12/58 6 tons. Sold to Hopkins England, Woburn Sands. 2nd half/65, resold to Jonallen. Singapore 8/66 as No. 1.

(g) NEW 20-4-29, £340, 60cm.

(h) ex L.B.L.R. 12/58 5 tons. Sold to Hopkins England 2nd half/65, resold to Jonallen 8/66 as No. 2.

(i)   del. 15-1-35 PLH, to GG 19-7-36.

(j)   NEW 31-12-36 £348, del. to DA.

(k)   del. 4-10-40 to Shelton Iron & Steel & Coal Co. Ltd., Stoke as 3½ ton.

(l)   NEW 8-12-36 £348 del. to Grovebury.

(m)   del. 12/36 to Sir R. McAlpine, sold to GG / 59.

(n)   del. 21-3-29 PLH, to GG 13-7-29.

(o)   del. 15-4-29 PLH, to GG 16-21-31.

(p)   del. 22-5-29 PLH, to Peter Land & Co. later. By 9/54 owned by G. Cohen who sold to GG. 60cm.

(q)   del. 18-10-33 to A.H.Worth, Fleet, Lincs. When purchased by GG fitted with cab having curved roof which it retains.

(r)   del. 19-9-44 to Ministry of Supply. Order of 57, 60cm gauge. Purchased from Bungey, Hayes.

(s)   del. 2-39 to Glasgow Corporation, Housing Dept., Robroyston Contract Order of 15. ex James N. Connell Ltd., Coatsbridge, via Bungey.

(t)   ex James N. Connell Ltd.

(u)   del. 30-8-37 to Glasgow Corporation Housing Dept., Pollock Scheme ex James N. Connell Ltd., via Bungey.

(v)   del. 11-11-36 DLH, sold 4-5-37 to Sir Lindsay Parkinson who sold to GG.

(w)   del. 5-10-36 Gloucestershire Tile & Sand Co.Ltd., Shurdington nr. Chelmsford. By 6/40 at H. Covington & Sons Ltd., Cremorne Wharf, Chelsea; ex **English** Clays, Lovering Pochin & Co.Ltd., Trinnick Mill, Cornwall, via Bungey.

(x)   del. 25-3-41 Sir R. McAlpine, Burghfield, Reading. Order of 15.

(y)   NEW, del. 19-11-36 £348.

(z)   del. DLH 25-1-36.

(aa) del. 17-5-27 PLH, reconditioned loco (from Heaton Park Reservoir) to GG 2-6-27, Note (b) probably refers to this locomotive.

(bb) presumed obtained SH, in 1950's, soon sold to Bungey.

(cc) del. France W.D.L.R. 2828 by 3/26 owned by Wilson, Kenmond & Marr Ltd.

(dd) rebuild of 369 del. 5/17 to France W.D.L.R. 1770 to GG from M.R. 13-1-26 £365 60cm.

(ee) del. 22-7-31 to Grovebury Road, reconditioned 60cm locomotive £200.

(ff) del. Sir R. McAlpine, sold to GG.

(gg) del. 21-8-34 Sir R. McAlpine, Kendoon Reservoir Contract, New Galloway, 3½ tons, sold to GG.

(hh) del. 27-2-36 Glasgow Corporation Housing Dept., Househill Wood Farm. First M.R. locomotive fitted with Vokes engine filter, previously Hirst filters used.

(ii) del. 31-7-39 Glasgow Corporation Housing Dept., Penille Contract. Order of 6.

(jj) del. 4-4-41 Sir R. McAlpine, Hayes depot (order of 15) sold to GG.

Note: this list represents a minimum, as seven unidentifiable locomotives were in the derelicts in 1958. Also, Garside's records of purchases from Motor Rail do not tie up with Motor Rail's sales. Possibly the sales were made from Petrol Loco Hirers Ltd., stock. This affects four locomotives, bought in February, March, July and Oftober 1925. This may well be among those for which no information is available in the list.

ROLLING STOCK

ROLLING STOCK

The L.B.L.R. owned no rolling stock, except perhaps one or two wagon underframes used to carry permanent way materials. All traffic was carried in wagons supplied by the quarries.

The commonest type of wagon used in the quarries and over the main line - an all steel side tipper - had a capacity of 1 cubic yard (25 cwt. of sand). The majority of these wagons were made by Robert Hudson & Co.Ltd., Leeds, although other manufacturers have been represented. Two "Excelsior" wagons by William Jones, London, were to be seen at Billington Road in 1968, while between the wars some Belgian and Japanese wagons (the latter with a rising sun embossed in the body ends) were used. Even amongst the Hudson wagons a number of varieties exist, but the general construction of all is very similar. The measurements given refer to a standard Hudson "Rugga" wagon (drawing 00/620374) but the figures are typical.

The underframe of 4" by 2" steel channel is oval, 6'2" long by 2'10" wide with almost semicircular ends; a 5" straight section covered by a steel buffing plate connects two quadrants. Modern wagons have the frame made of a single piece of channel whose flanges face inwards. Earlier wagons had the frame constructed from four pieces of channel. Two straight pieces with the flange facing outwards were rivetted to curved end pieces with inward facing flanges. The sides are kept apart by a central crossbar of light angle. The manufacturers recommend that where locomotive haulage is employed the wagons should have underframes with a central channel longitudinal to take heavier buffing and drawbar stresses, but wagons with and without this longitudinal are used, both within the quarries and over the L.B.LR.

The wheelbase is 22" or 27". Early axleboxes had plain bearings; now ball or roller bearings are the rule. Springing is by rubber blocks. The 4 or 6 hole disc

ROLLING STOCK

wheels are mainly 12" in diameter, but most of the wagons used by Arnold's for main line traffic are fitted with 16" wheels for improved riding. Large and small wheeled wagons tend not to be marshalled together. Some of their small wheel wagons with good bearings are branded "Main line". ('Garside's once had a few large-wheeled wagons for the main line). A pocket on top of the end frames take the oval single link couplings which are secured by vertical pins.

Some of Garside's wagons were brake-fitted to help control trains hauled by light locomotives on the steep gradients and to control wagons running into the pits. The underframe was lengthened to 7'6" to accommodate the vertical column for a screw brake which acted on all wheels. A steel floor was provided by the column for the brakeman. The end overhang was unequal - about 3' at the brake end and 2'3" at the other. The wagon brakes were not often used and by the late 1960's had been removed, although the long asymetric underframes were unaltered. Several may still be found at Double Arches.

The steel V section body is 2'4" in depth and 4'7" wide inside, 5'0" overall. The ends taper slightly to aid discharge, so that the internal length is 4'6" (top) and 4'0" (bottom). Wagons used mainly to carry wet sand from the washers (or Grovebury dredger sand hoppers) have a row of holes drilled along the base of the body to allow water to drain. The wagon is $3'9\frac{1}{4}"$ high. Two pressed steel standards at each end carry a length of horizontal angle fitted with end stops. A piece of curved angle on the body ends rolls along this horizontal track enabling the body to tip to either side. A low pin in the centre of the track engages in a hole in the curved channel to locate the body. A simple pivoted catch underneath the body either at one end or in the centre automatically locks it in an upright position. It tips away from the operator.

ROLLING STOCK

Early "Victory" wagons had curved channel steel standards. The body ends carried a piece of straight angle which rolled on the standard. A lever fastened to the standard engaged with the body angle to prevent tipping.

A type built until about 1938 had one piece pressed steel stands and bridle catches. The catch was an inverted U of iron, pivoted to the frame side members which curved over the body carrying angle, and so prevented it from rising. To tip the body the catch was moved outwards to clear the angle.

At least one wagon, probably secondhand, in the early days had lifting eyes at each corner of the body, and four feet underneath, so that the body could be lifted off by a crane. It could not be tipped while in the crane slings.

Arnold's owned about 400 tipping wagons in 1961 and Garside's about 250. Wagons are painted brown and grey respectively. They are numbered, Arnold's favouring the stand and Garside's the body end, for the white figures. Arnold's also paint the date of the last overhaul on the frame. Wagons weigh about 9 cwt. empty.

When overburden was removed by hand some end tipping wagons were used. They were basically a standard side tipper, complete, mounted at right angles on an additional chassis. The new chassis might be full length, in which case wagons could be coupled but on discharge the load did not clear the frame ends very well; or one curved end could be removed so that discharge was unobstructed but the wagon could be coupled only to the rear of a train; or both curved ends could be removed which gave clear discharge either end but made coupling impossible. One of Garside's wagons of the last type survives in the hands of the Iron Horse Preservation Society.

Some sand was dried and loaded into sacks. To carry the sacks the short distance from their Billington Road plant to the main line sidings, Garside's had four

ROLLING STOCK

bogie flat wagons. The first was constructed in 1924 and the others as required. The bogies were standard tipping wagons underframes with a piece of heavy channel steel for the bolster. The body frame was made of similar 8"by 4"channel (or in one case wood) with a transverse piece near each end to carry the bogie pivot. The body was 13'6" long by 4' wide and the wooden floor was 2'9" above the rails. The overall length was about 15' 6". These wagons had a tare weight of about 18 or 19 cwt, and could carry 4 tons.

For the longer haul from their plant at Double Arches, Arnold's used eight ex-W.D.L.R. Class D bogie open wagons built by Gloucester Carriage & Wagon Co. Ltd. These had 3-plank wooden bodies approximately 17'6" by 5' by 2' on a steel frame. Each side had two drop doors, separated only by a removable steel pillar. The bogies extended a short distance beyond the body because originally they had been fitted with screw brakes operated from a pillar mounted on each bogie. The wheels ran in plain bearings in axle boxes supported by laminated springs. The wagons had a tare weight of about 2 tons 8 cwt. and could carry about 6 tons of bagged sand. The standard W.D.L.R. coupling was a single link held vertically in the buffer coupling. So that these wagons could be coupled to others which used a single horizontal link the buffers of the bogie wagons were turned through 90 degrees in their sockets. This obviated coupling difficulties but greatly reduced the effective width of the buffer face and so facilitated buffer locking. To prevent this, these wagons were coupled by "treble links", two links joined by a ring, as used on the locomotives. The wagons were numbered 1-8 inclusive. In June 1969, following the closure of the southern half of the railway the 7 survivors were offered for sale.

There are several special wagons. One of Arnold's class D wagons was converted into a bogie flat in the early 1960's to carry bulldozers between the quarries and Billington Road shops. The sides of the frame were extended slightly at each end to carry a transverse bar, $1\frac{1}{2}$" in diameter. Ramps, 6'9" lengths of 8"by 4", I girder with

ROLLING STOCK

wooden filling in one channel, could be hooked onto this bar so that bulldozers could be driven onto the wagon. The first time the wagon was used the floor was so badly damaged by the weight of its load when the wagon twisted slightly on rounding a curve that it had to be replaced with thicker planks.

Garside's have an unusual Hudson open wagon with a body 5'2" by 4' by 2'. An old type underframe carries a length of 4" deep channel along the top of each frame side member. This channel supports a wooden floor of 2" planks. Steel ends are bolted to the floor and are braced by vertical angle irons from the underframe. The steel sides lift out. The base has two pegs which engage in holes in two metal strips projecting beyond the floor, and the top corners are secured by conventional pegs and cotters. The wagon was used to carry excavator spares.

Other wagons are conversions using standard wagon underframes as the starting point. Both companies have well-equipped workshops for repair work and can easily undertake rebuilding. Garside's bogie wagons have already been mentioned, and a pair of 6 wheel flats were built for Grovebury (q.v.) using wagon and locomotive spares. They also fitted several underframes with a rack to hold an oil drum, and fitted a hand pump at one end, so that excavators could be easily refuelled in the quarries. There were also two tank wagons, one for oil at Grovebury and the other for carrying weed killer, used in conjunction with a spray wagon. This had a pump driven off one axle and fed a transverse horizontal perforated pipe.

Arnold and Garside both had a similar wagon to carry small road trailers into the quarries. Wide channel side members were welded on the outside of the wagon frame to form troughs for the wheels. Arnold's vehicle carried an air compressor for rock drills at Chamberlain's Barn and Garside's held a portable electric welder kept in their Double Arches navvy repair shops.

The most interesting of the homemade vehicles was Garside's crane constructed around 1940. The wheelbase

ROLLING STOCK

of the underframe was increased to 33". A length of standard gauge bullhead rail was mounted at each end of the frame to give extra stability. A simple girder frame formed the base of the pivot. The crane body turned on four rollers running on this frame. The body – 5' long by 2' wide – was built on a frame of 4" by 2" channel. A pair of triangular frames carried the geared drive to the drum and a band brake. A box at the back held the winding handle and scrap for balance weight. The jib was 9' long over centres and was held by a wire loop fastened to the back of the body so that the hook had a fixed radius of about 7'6". The crane spent most of its life at Billington Road, and is now owned by the Iron Horse group.

Garside's have 3 buffer-beam snowploughs made by Hudson's in the late twenties – i.e. when locomotives were becoming common in the quarries. The plough is a V of two $\frac{1}{8}$" steel plates, 1'4" deep at the nose and 1'6" deep at the wings. A central strut about 5' long runs from the nose to end in the locomotive coupling casting where it is secured by the coupling pin. A transverse member 4'5" long holds the wings apart. One plough has been used at Grovebury, one at Billington Road and one at Double Arches where two are currently stored.

QUARRY WAGON STOCK

J. ARNOLD / G. GARSIDE

1961	BR	CB	9A	DA	Total.	BR	MH	DA	Gr	Total
Tippers: 16"whl.	48	8	19	80	155	—	—	—	—	—
12" "	32	80	20	121	253	65	40	120	43	268
brake fitted.	—	—	—	—		4	—	2	—	6
Bogie open	—	—	—	8	8	4	—	—	—	—
" flat	—	—	—	—	—	—	—	—	—	4
oil drum/tank	—	—	—	—	—	—	1	3	—	4
Misc.	4	—	—	2	6	2	—	3	3	8

1968										
Tippers: 16"whl.	44	3	16	98	141	—	—	—	—	—
12" "	2	73	17	56	148	—	9	160	9	178
Bogie open	—	—	—	7+	7	—	—	3	1	4
" flat	—	1	—	1	1	—	—	2	—	3
Oildrum/tank	—	1	—	1	2	—	1	3	—	3
Misc.	3	2	—	2	7	—	—	—	—	—
Horses	5*	1	2/3	5	13/14	1/2	—	3	4	8/9

Note1. Figures include wagons needing repair and stray underframes, so that totals are approximate. However, they give an indication of the size and distribution of stock.
2. Horse allocations refer to the period before the quarries had many locomotives.

* Washer (low level)1, Transhipment Ramp 4.

\+ **Sold July 1969 to Festiniog Rly.** 4 Iron Horse 3

IRON HORSE PRESERVATION SOCIETY

IRON HORSE PRESERVATION SOCIETY

In recent years increasing numbers of railway enthusiasts have become involved in preservation. Narrow gauge industrial locomotives have become popular due to their small size and weight (one enthusiast has a 2' gauge 0-4-0ST, wagons and track which he takes to traction engine rallies, an operation not to be contemplated with even the smallest standard gauge stock.) On January 1967 a number of enthusiasts met at St. Albans to "discuss the possibilities of a narrow gauge railway somewhere in the vicinity", but as no progress was made L.C.Brooks and B.J.Harris decided to proceed alone.

They then attempted to purchase materials from the L.B.L.R., only to be told Arnold's had no intention of closing it. On January 1967, Mr.J.Arnold granted permission for them to run trains over the L.B.L.R. at weekends. In the following March they obtained permission to use sheds im Garside's disused Billington Road depot for storage and repairs.

Soon after, four Simplex diesel locomotives were purchased from St. Albans Sand & Gravel Co., Smallford Pits, and taken to Leighton Buzzard. Wagons from the Colne Valley Water Board line at Watford followed. Volunteers began to work although the Society was not officially formed until October. Until now Brooks and Harris had supplied the capital required. It was the "wish of the principals to run the railway as an American style narrow gauge operation". So far the American aspect does not appear to have reached beyond publicity. A site was levelled beside Page's Park loops for a depot (929242), and in the spring of 1968 the green corrugated iron shed (30' by 15') was erected. It contains two tracks, laid on longitudinal BR sleepers which thus form a shallow pit. A bay 10' square provides extra storage space.

To provide space for the shed the Society lifted Garside's loop. Between Arnold's loop and the main line

IRON HORSE PRESERVATION SOCIETY

a low ash platform edged with sleepers was made. Sidings have been laid behind the shed, and members have done some track repairs on the L.B.L.R. mainline.

The first public passenger trains over the L.B.L.R. were organised by the society for Sunday, March 3rd, 1968. The first, leaving Page's Park at 10.30 comprised three of Arnold's bogie open wagons hauled by IHRR 1. Threequarters of an hour were allowed at Arnold's Double Arches quarry for the train to turn round and for visitors to explore. On the return trip it crossed a second train, two bogie opens hauled by Arnold's No. 7. In the afternoon the first train made the third journey. Mr.Harris said "This is the first of what we hope will be many trips and next year we should get going seriously." By the time this appears the prophesy should be verified, although the line will presumably have to be inspected by the MoT first. With the impending closure of the main line sidings at Billington Road, the Society has agreed to take full responsibility for, and have sole use of, the L.B.L.R. as far as Vandyke Road, as from May 1st, 1969.

The Society owns four diesel locomotives. None were working order when purchased, but after much hard work and cannibalisation, two are now restored. Three steam locomotives are on loan. "Chaloner," a vertical boilered vertical cylindered locomotive which formerly worked at the Pen-yr-Orsedd slate quarry, near Caernarvon, "Pixie," formerly owned by Devon County Council and "The Doll," one of three locomotives built for an Oxfordshire ironstone quarry and later used at Springvale Furnaces, Bilston. "Chaloner" had to be regauged after arrival as it was found that one pair of wheels was 1'11½" gauge and the other pair 1'10½". Locomotives of this type were once very common in quarries in North Wales. Also on loan is "Rail Taxi", a BMW Isetta bubble car converted to run on rails by fitting flanged wheels to the rear hubs and a bogie under the front end. It is said to be capable of 50mp.h. plus. In February, 1969, "Rail Taxi" was transferred to B.Goodchild's railway near Leamington Spa.

IRON HORSE PRESERVATION SOCIETY

The working Simplex locomotives have a blue livery with black frames, yellow chevrons on the ends and red handrails. "Chaloner" has black boiler and frames, green tanks and bunker with black and white lining. The ploughs at each end are red, as is the livery on the boiler bands. Rail Taxi was blue and white with black lining. All carry the Society's crest - a white prancing horse on a red ground encircled by Iron Horse Railroad Leighton Buzzard on a golden outer circle.

Passenger rolling stock comprises a blue 16 seat open coach, built on an ex WD bogie open frame, with screw brakes on one bogie; and two skip frames, one blue and the other grey, fitted with longitudinal seats. Other rolling stock comprises two Hudson side tipping wagons, in red oxide with white lettering - Nos. 101 and 102 - a Hudson end tipping wagon, a flat trolley and Garside's crane.

1	4wD	MR	5608/1931	(a)
2	4wD	MR	5613/1931	(b)
3	4wD	MR	5612/1931	(c)
4	4wD	MR	5875/1935	(d)
Chaloner	0-4-0T VB	De Winton	/1877	(e)
Pixie	0-4-0ST	Kerr Stuart	4260/1922	(f)
The Doll	0-6-0T	Andrew Barclay	1641/1919	(g)
Rail Taxi	4-2-0P	BMW Isetta		(h)

(a) 20HP 4 ton, plate frames Wm.Moss & Sons Ltd., Loughborough 2/7/31. By 2/37 owned by St. Albans S&G. R9.

(b) 20HP 4 ton plate frames Winfields Ltd., Birmingham but delivered to Hussey Egan & Pickmere Ltd., Prittlewell Sewage Works, Southend 1/9/31. By 9/37 owned by St.Albans S&G: R7.

(c) As (b) but dispatched 20/8/31. Spares to St. Albans S&G 6/37. R8;

IRON HORSE PRESERVATION SOCIETY

(d) 20/28HP 2½ ton channel frames PLH 25/3/35.
After short period of hire, to St. Albans S&G,
Meadgate Farm, Nazing, Essex. 28/6/35 R3.

(e) New to Penybryn Slate Quarry Co., Nantlle, to
Pen-yr-Orsedd Slate Quarry Co.Ltd., Nantlle
4/1892. to Kings Langley for preservations 9/60,
to LB 22/6/68.

(f) Wren class bought by Devon C.C. 1929 from T.W.Ward
Ltd., Grays, Essex, to Industrial Locomotive
Society. Berkhampsted for preservation 7/57, to
LB 10/12/68

(g) New to Alfred Hickman Ltd., Sydenham Pits, near
Kings Sutton, Oxfordshire. To Stewarts & Lloyds
Ltd., Springvale Furnaces, Bilston, about 1926.
Sold for private preservation at Coventry 1960,
to Bressingham Gardens, Norfolk, 1966, to LB 1969.

(h) On loan by R.Morris, Longfield, Kent; to B.Goodchild
Leamington Spa 2/69.

QUARRY SYSTEMS NOT CONNECTED TO L.B.L.R.

QUARRY SYSTEMS NOT CONNECTED TO L.B.L.R.

Bedford Silica Sand Mines Ltd., Heath & Reach.

The adjoining hamlets of Heath and Reach, about two miles north of Leighton Buzzard on the main road to Woburn (A418) were made into a single civil parish in the last quarter of the nineteenth century. Of the several sand pits in the vicinity of the village the largest is situated to the east of a lane linking Heath and Reach. (926284)

This pit has been in production since around the turn of the century. The 1914 Bedfordshire Trade Directory gives "Christopher Claridge, Sand Merchant, Heath & Reach" and in the 1928 edition this changes to the "Waste Recovery Syndicate Ltd., Silver Sand Merchants". The Syndicate did not operate for long. The present company, Bedford Silica Sand Mines Ltd. - "speciality dried and graded sand" - appears in the 1931 edition. This company was the first to dry and grade sand in this country, the necessary plant being opened in 1928. The railway was laid at about the same time, no doubt to deal with the increased output of the new process and to ease transport from quarry to works. Previously horses and carts had been loaded at the quarry face.

The track layout was in the form of a Y, with the drying plant at the centre. The two arms led eastwards to the sand pit, although the northern one had been disused since about 1950, and lifted in 1953 - a pair of Motor Rail dumpers being used to carry the sand instead. The track comprising the stem of the Y ran westwards for about 200 yards downhill, but on an embankment of gradually increasing height, to a wooden tipping dock beside the main road. This section was abandoned in 1939 when a **brick** tipping dock was built beside the works. The embankment remains, overgrown, at the bottom of the gardens of a housing estate while the tip itself was removed and a house, 32 Birds Hill, built on the site in 1960.

QUARRY SYSTEMS NOT CONNECTED TO L.B.L.R.

A railway was also used to transport overburden until 1936 when it and the barrow runs were replaced by a Smith (Rodley) excavator and dumper. A small portable screen fed by a bucket conveyor was installed in the pit at about the same time. Men now shovelled sand into the conveyor hopper instead of direct into the wagon.

The railway was operated by gravity with 2 or 3 horses to haul wagons back uphill. Trains comprised 3 or 4 wagons with the leading one fitted with a platform at one end on which a man rode to apply the screw brake when necessary. About 1 in 4 of the Hudson side tipping wagons were brake fitted. After the screen was introduced, a loop was laid in the pit with the delivery chute from the screen above one track. Empty wagons were left on the other and pushed by hand under the chute as required. Previously there had been a number of short spurs to the different parts of the face, reached by points or by turning plates. The points might be of conventional design or made to fit <u>on top</u> of permanent track so that a temporary spur could be laid in, without cutting the main line at the junction.

The horses were replaced by two 2½ ton Motor Rail 20/28 HP diesel locomotives, Nos. 8588 and 8592, both built in 1942. No. 8592 was bought on 12th June, 1946 from Diesel Locomotive Hirers, Bedford. It was cabless. The second locomotive was obtained on 25th March, 1948. It had been sold in the first place to C.T.Olley, Grove Farm Sand Pits, South Ockendon, Essex, then resold to Motor Rail on 10th December, 1947, who had fully reconditioned it. A cab was fitted.

Trains were hauled from the pit and empties pushed back. Loads of screened but undried sand were pushed onto the tipping dock. In 1961 there were 58 Hudson wagons (not all in use). The track was 20 and 30 lb. rail, laid on halves of ex-BR sleepers. The railway then was nearing the end of its life, since it could not supply the elevated high-capacity bunker necessary if the work was to increase its output and efficiency.

QUARRY SYSTEMS NOT CONNECTED TO L.B.L.R.

A new 100 ton bunker was built at the works, fed by a conveyor from a 30 ton bunker in the pit. This bunker was loaded direct. The plant was then able to dry 20 tons of sand per hour, continuously, which had been impossible with rail supply of sand. A train of 4 wagons had brought only 6 tons at a time and little stock was held in the existing hoppers. The new conveyor was brought into use in May 1963, and the railway finished. The locomotives were sold in August, 1963, to Flettons Ltd., Kings Dike, Cambridgeshire, and the rest went for scrap.

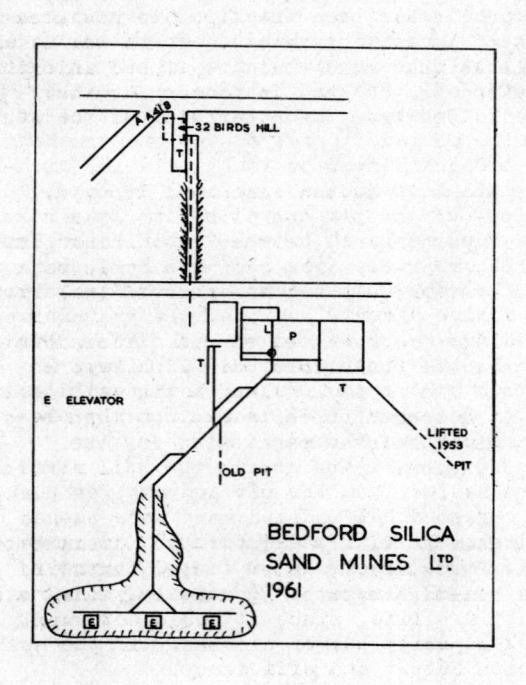

BEDFORD SILICA SAND MINES LTD 1961

QUARRY SYSTEMS NOT CONNECTED TO L.B.L.R.

H.G.Brown, Kings Farm, Leighton Buzzard.

H.G.Brown, Kings Farm, Stanbridge Road, Leighton, was a timber and sand haulier. He cut the original sleepers for the L.B.L.R. in 1919. Around 1920 he began to dig sand from a pit between his farm and Page's Park (929246). A sleeper road was laid into the pit so that carts could be loaded at the face. About 1926 this was replaced by a light 2' gauge railway.

The railway climbed out of the pit onto an embankment and terminated on a gantry of heavy timbers with a sleeper floor. The gantry was built beside the farm road and against a wall of the shed where Brown kept his steam wagons. (He had been the first to use steam wagons to haul sand in this district). There was a large door in this wall so that sand could be tipped into the wagon while it stood in the shed in the dry. Other men's carts were loaded from the other side of the gantry in the open.

There were about 20 Hudson wagons of 1 cu.yd. capacity hauled out of the pit one at a time by a horse. Sidings in the pit were mainly reached by turning plates instead of points. Six or eight men were employed digging sand and overburden which was removed in barrows.

In 1935 or 1936 Brown wanted to run a line down to Grovebury sidings so he could load main line wagons direct. He made a gap about 20' long in the wall beside Billington Road (the houses opposite had not then been built) but the Council refused permission for the necessary level crossing. The gap in the wall remains today.

One of his employees, "Fatty" Orchard, is remembered for his ability to transfer $1\frac{1}{2}$ cu.yd (say 37 cwt.) of sand from a cart to a railway wagon in $3\frac{1}{2}$ minutes, using a No.4 pan coke shovel.

QUARRY SYSTEMS NOT CONNECTED TO L.B.L.R.

In 1947 the quarry was let to Henry Winfield and the railway finished. Winfield laid a road of "conks" – sand rejects from the screens – for his lorries which were loaded through a $\frac{1}{2}$" vibrating screen, direct by a 10 RB excavator. This pit's sand had few rejects, so conks for the road were brought from Winfield's other pit at Heath and Reach. The pit closed in 1963.

A short 2' gauge line was laid between the pit and Page's Park by H.Paul. He started a small unauthorised breeze block making plant beside the park in about 1950. The railway was laid to take the blocks to concrete drying areas nearby. The wagons were tipping wagon frames with wooden floors, and were pushed by hand. However, as the plant had been built without planning permission the Council in 1954 or 55 ordered it to be closed. The shed and a length of track set in concrete remain in the back of C.Nash's yard and can be seen from the Park.

QUARRY SYSTEMS NOT CONNECTED TO L.B.L.R.

Clay Cross Coal Depot, Linslade (913243)

The October 1967 "Bulletin of Industrial Archaeology in CBA Group 9" gives a short paragraph to this tramway "which appeared on the Dunstable and London and Birmingham Railway plans of 1844. The tramway is shown as running from 'Clay Cross Coal Depot' just south of Linslade to the Grand Junction Canal, apparently not joining the London and Birmingham Railway and crossing the Dunstable line at an acute angle. The depot, once had a limekiln.. Geoffrey Webb of Bedford has found no surviving traces of the line."

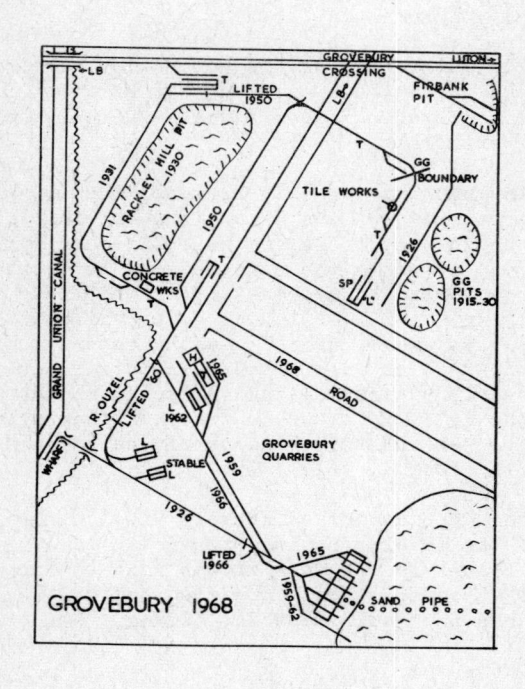

QUARRY SYSTEMS NOT CONNECTED TO L.B.L.R.

Grovebury Quarries

The area of Grovebury, about a mile south of Leighton Buzzard, contains a number of sand pits. All but one are now disused.

The Leighton Buzzard Sand Co. Ltd., worked a quarry, Firbank Pit (924240) to the south of the L.N.W.R. Dunstable branch east of Grovebury Crossing. A standard gauge siding ran from near the level-crossing along the southern edge of the pit. In 1925 this siding divided into a long loop, and a siding from this reached almost to the old pit called Spinney Pool (927237) which Joseph Arnold had opened about 1880. In 1937 the loop had been partly lifted so that only a pair of sidings remained, spanned by a building, probably a hopper for sand pumped from the now flooded pit. The pit gradually fell out of use and is now a tip. The sidings, extended, remain, and are used for storing BR Engineers wagons. Shunting was presumably done by main line locomotives or horses.

Close to the northern end of the loop, the 1925 map shows a separate single line of railway running about 200 yards south west to serve two small sand pits (923237 and 923239) close to the lane leading to Grovebury Farm. George Garside owned these pits which had been opened early in the 1914-1918 war. The railway was presumably narrow gauge, probably using materials from Billington Road which was on the verge of exhaustion. These pits were early victims of the recession in the 'Twenties' and the 1937 map shows no trace of the railway. Nothing can now be seen of this line or of the (presumed) tipping dock at the loop.

Joseph Arnold & Sons worked Rackley Hill quarry (920241) west of the level crossing, for some years until sometime in the 1914-1918 war when it was sold to George Garside, probably because it only produced building sand and Billington Road pit could meet the demand.

QUARRY SYSTEMS NOT CONNECTED TO L.B.L.R.

Rackley Hill became Garside's main quarry. In the 1920's - and probably before - a short line ran from hoppers at the northern end of the pit to an elevated wooden tipping dock beside a standard gauge siding (920242). The pit was flooded and sand was dug by a steam grab mounted on a pontoon. When full the ensemble moved across to the hoppers and the sand was transferred. The level fell considerably when the water table was lowered following the construction of new water boreholes in the neighbourhood. The pit was worked out in the early 1930's and production then centred on Grovebury quarry nearby.(919234)

Grovebury was opened in the mid 1920's. A short railway is shown on the 1926 map. The line ran from the pit northwest to the River Ouzel where it divided. One branch crossed the river by a sleeper bridge, and then continued on a low embankment for about 100 yards to the Grand Union Canal where it turned sharp left on to a private wharf where the track was set in concrete. Sand was sent by boat to London, and while in recent years traffic dropped to about a boat a fortnight, the last boat was not loaded until October 1965. The right-hand branch curved sharply through a right angle and ran straight for about 200 yards to a tip near the divergence of the 1953 cut off.

Leighton Buzzard Tiles Ltd. opened their works (921240) in 1930/31 close to Grovebury Crossing. To supply the works with sand a long extension was built. The tile works branch curved sharply left from the former route to run round the western and northern sides of Rackley Hill quarry. It made connection with the older line to the L.N.W.R. tipping dock. Leaving the ramp to this dock the track continued towards Grovebury Road which it crossed by an ungated crossing and ran along the side of the tileworks for 200 yards. Here sand was tipped. The distance from the sand hoppers, which had been built by the pit in 1926 when a sand pump was installed, to the crossing was about 1200 yards.

QUARRY SYSTEMS NOT CONNECTED TO L.B.L.R.

By 1937 a corrugated iron locomotive shed had been built by the old wooden stables at the canal branch junction, and later the stables were used for locomotives as well. Also, a branch had been built continuing from the original line beside the quarry approach road to a tipping dock just inside the entrance. The dock had a wooden floor supported by steel girders resting on brick piers. It was about 600 yards from the hoppers and about 200 yards from the tileworks crossing, so the extent of the deviation round the old pit can be seen. The overburden was removed by rail at this time, the 1937 map showing a line running from near the locomotive shed round the eastern tip of the quarry.

After World War II the system was progressively modified. The first alteration was the construction of a short line from near the tipping dock by the entrance to join the tileworks line by the level crossing. This was brought into use in about 1950 and the route round the old pit fell into disuse as little sand was sent away by rail. The railway tip was last used in April 1965. However, the first few yards of this line had received a new lease of life in June 1964 when Readymix Transite Ltd. built a mix concrete plant beside it, near the site of a loop shown on the 1937 map but gone by 1958. A sunken hopper fed sand to a conveyor which took it up into the plant.

To reach either the lorry bay or the tile works, trains had to travel round two sides of a triangle from the original terminus. A deviation was built along the hypotenuse and opened in December 1958. There was no connection by the original terminus but the track thence to the locomotive sheds remained until about 1964. The section from the sheds to the hoppers was needed to give access to the former and to the wharf. This part of the line was laid in a wooded cutting and was very picturesque - as also was the line to the west of the old pit which ran through a wood - but since closure of the wharf in October 1965 and the erection of a new

QUARRY SYSTEMS NOT CONNECTED TO L.B.L.R.

locomotive shed, this section has become very overgrown.

A spur was laid off the hypotenuse to a hopper sunk in the ground. A conveyor took the sand from this hopper to an elevated hopper from which lorries were loaded. A small screen was incorporated and grit fell off into a wagon on a short siding. With the introduction of this hopper the lorry tip by the gate lost its importance, but was used to supply a stockpile in slack periods as a reserve against breakdown. In mid 1965 a second line was laid beside the hopper. At both positions a corrugated iron shelter was provided for the man tipping the wagons.

With the deviation in use the locomotive sheds were remote. They were also small and inconvenient. Accordingly in the latter part of 1962 a new wooden shed - 46' by 18'9" - with a single road to take four locomotives was built beside the new tipping hopper. The floor of the new shed was still only sand with no pit. The old sheds remain in the wilderness, the locomotive shed with faded maroon paint over its corrugated iron.

The tileworks line was extended by Eastwoods Ltd. (who had taken over Leighton Buzzard Tiles (1934) Ltd. in 1934 as a subsidiary) sometime after 1937 to serve more tipping positions at the back of the factory. The line beside the works remained Garside's as it was a potential source of access to Garside's old quarries behind the works if they should ever be reopened. Also Garside's bought more land south of the farm since 1945. Eastwood's extension was about 150 yards long and terminated on a gantry about 7' high across a brick sandstore. Sand was kept here in case the supply from Garside's should ever be interrupted. Following a period when Garside's had a number of locomotives out of action, in 1957 the company acquired about a dozen wagons, and an Orenstein & Koppel diesel locomotive so that they could collect their own sand. One informant said that

QUARRY SYSTEMS NOT CONNECTED TO L.B.L.R.

a locomotive from Eastwood's Barrington Cement Works was borrowed for a year before the Koppel came.

Further emergencies rarely arose. The locomotive was used but once and stood idle at the end of the gantry. In 1961 I found a bird's nest in its radiator ! Redland Tiles Ltd. took control on April 1st 1963. The new owners reorganised the works. It was found more convenient to have sand delivered by lorry and the railway became redundant from May 1964. The track was soon lifted (apart from the rails in the level crossing) and by late 1965 a factory yard was being extended over the old track bed. The locomotive disappeared, presumably for scrap, in 1964.

The track between the junction of the lines to the tips and the readymix plant, and the pit sand hoppers was doubled in 1966. The connection to the old sheds was taken out at this time. This was the last modification. The existing plant involved double handling and so new equipment which loaded lorries direct was installed on the eastern edge of the pit. These came into use at the end of May 1968. The stocks of sand in the old hoppers were sufficient to supply Readymix for the first week of June and then the railway closed. The equipment subsequently had painted on it "Keep" (to be sent to Double Arches later) or "Scrap".

Horses were used initially. The first locomotives came second-hand in the mid 1920's. They were both made by Austro-Daimler. In one the driver sat sideways in the midde, and in the other a wagon was incorporated as part of the locomotive, Later it was de-engined and used as a "Bogie". Further locomotives were standard Motor Rail designs. A reconditioned locomotive was supplied in July 1931 (for the tileworks traffic) and two new ones in December 1936, but these were probably not the first Motor Rail locomotives to work here. There were three locomotives in 1945. In the 1950's and 1960's the stock was usually four locomotives, with occasional transfers to and from Double Arches for heavy repairs. For details of Garside's locomotives used on

QUARRY SYSTEMS NOT CONNECTED TO L.B.L.R.

this system see the table at the end of Chapter 8.

Eastwood's Ltd. locomotives

4 wD   on loan from Barrington Cement Works c 1956/57.

4 wD   Orenstein & Koppel 4105 ex Eastwood's Bobbing ClayPits, Kent c/57 scrapped 1964.

    Garside's locomotives had the standard green livery with white uppers to the cab. They all ran with the bonnet facing the pit. Eastwood's diesel had black bonnet and frames, and its well constructed home-made cab painted cream with "Eastwood's Limited" in black on the closed side. This locomotive faced the other way.

    The first horse drawn wagons had been combined with side and end tippers, later Hudson side tippers holding 1 cu.yd. (25 cwt.) of sand had been used. In the 1960's there were about 50 to 60 wagons on the side although only about 36 were in use. At the closure 35 wagons were marked "keep" (later sent to Double Arches) and the rest "scrap". Eastwood's had 12 similar wagons, up to 4 stored by the locomotive and the remainder scattered beside the track at the back of the works. Their sand was brought in Garside's wagons.

    In addition to the skips there were 4 other wagons. One was a skip frame carrying a rectangular oil tank. It was kept behind the sand hoppers. Two wagons were robust home-made 6 wheel flats, each with a central pivoted bolster. The frame was channel steel (6'1" long over buffers, 3'1" wide (3'8" bufferbeam) 1'2½" to the top of the frame from the rail, 2'3" wheel base, 10' by 1' bolster). It was well braced, and had ex petrol locomotive buffers at each end. The wheels and bearings were standard Hudson products. The centre pair of wheels was flangeless. These two wagons were made to carry a new pontoon weighing 21 tons down to the lake. It arrived on March 12th, 1959. A siding was laid down to the water by the sand hoppers

QUARRY SYSTEMS NOT CONNECTED TO L.B.L.R.

and the loaded wagons, coupled by a long steel bar, pushed down it into the water. When the pontoon had floated clear the wagons were recovered and dumped on the old line close to the original terminus. The wagons were used in late 1968 to carry the dismantled hoppers up to the new locomotive shed. The fourth wagon was a bogie flat which came from Billington Road around 1965 and used for carrying stores etc. The 6 wheel and bogie wagons, also the locomotives, were transferred to Double Arches in the early months of 1969.

Three locomotives were normally in daily use. All trains were hauled from the pit and the empties pushed back. Trains were of 6 wagons, making a load of about 7½ tons of sand. The normal weekly quarry output was about 3500 tons. The tileworks took normally 16 train loads a day and the mixconcrete plant up to 12 a day. Trains passed either at the sand hopper, one leaving as soon as one arrived, or at the tip junction. For winter operation a small buffer beam snowplough was kept by the old locomotive shed but this had gone to Double Arches by the 1960's, presumably in the hard winter of 1962/63.

QUARRY SYSTEMS NOT CONNECTED TO L.B.L.R.

Leighton Buzzard Brick Co.Ltd.

(a) Ledburn Road Pit, Leighton Buzzard:

The Leighton Buzzard Brick Co.Ltd. opened its pit at Ledburn Road (914236) a mile southwest of Leighton in 1925. At first the sand was barrowed to carts or occasionally lorries, which carried it to the railway at Wing Sidings. Most customers were Luton firms and sand was sent by rail, via Bedford, to the M.R.sidings at Luton.

A 2' gauge railway was introduced in 1927 as the pit became deeper. There were 6 or 8 Hudson side tipping wagons, capacity 1 cu.yd. They were hauled singly up to the loading stage by the road, by a steam winch.

Later the winch was replaced by a more powerful petrol driven one capable of hauling three wagons at a time up to the tip under which the lorries backed for loading. A pony was used to haul the trucks from the several faces to the foot of the incline, where they were coupled into trains of three and hooked to the cable. The incline eventually had a length of about 200 yards. There were now about a dozen wagons in use, none brake fitted.

The sand was dug by hand, the men often standing on ledges cut in the pit face which averaged 35' high. When pockets of unclean sand were struck the sand was thrown through a nearly vertical screen mounted either beside the wagon or on the ledge. In 1935 the Company obtained its first excavator - a Priestman. A sleeper road was laid down and lorries loaded direct at the face. The railway was then used to remove overburden. It eventually fell out of use and most of the equipment was sent to the Watling Street (Potsgrove) pit when it opened in 1944. (A sleeper road was laid down from a new entrance at the other end of the field and lorries loaded direct at the face.) Ledburn Road pit continues to be worked. Sand is now pumped into hoppers which load lorries directly from workings across the road, beside the railway.

QUARRY SYSTEMS NOT CONNECTED TO L.B.L.R.

(b) **Potsgrove Pit**:

On the south side of the Watling Street (A5) beside a garage and the junction with a lane leading to the hamlet of Potsgrove, are the sand quarries of the Leighton Buzzard Brick Co.Ltd. (939298). These quarries, adjoining recent extensions to Garside's Double Arches workings, were opened in September, 1944.

A 2' gauge railway was built to carry the sand up out of the pit. Some of the materials came from the Ledburn Road pit system. The track was laid in the concrete at the edge of a steep road leading to the pit. At the top of the gradient the line curved north across this road to run parallel with the A5. A trailing point gave access to a level siding which was laid on girders supported by brick pillars beside the steep road. Here lorries were loaded.

The railway had but a short life. Arthur Wells visited the quarry in November, 1945, to see the steam locomotive "Haig" which stood on a short length of track just inside the entrance. He does not remember having seen any more track or railway equipment, so presumably the line was then still being built. It operated for only about a year and closed in 1946. The haul was too short, the gradients severe and so it was found more economical to lay a sleeper road in the pit thus enabling lorries to be loaded directly by the excavator.

Following the closure the track was lifted and all the equipment dumped on the waste ground near the A5, where most of it remained until 1952. The two Ruston & Hornsby diesel locomotives were sold. (The Motor Rail locomotive had been sold in 1949). "Haig" and the wagons were scrapped. All that remained in 1961 was a short length of track embedded in the concrete road and a battered Hudson wagon frame with an oil drum chained to it.

QUARRY SYSTEMS NOT CONNECTED TO L.B.L.R.

Despite its small size and short life the railway had no less than four locomotives. The first was "Haig", a Kerr Stuart "Wren" class 0-4-0ST, No. 3105, built in 1918 for the construction of Kidbrooke Aerodrome, London. "Haig" spent about 25 years at Kidbrooke before being sold to the Leighton Buzzard Brick Co.Ltd. After the arrival of the diesel locomotives, "Haig" did little work. In February, 1952, G.P.Roberts found the boiler lying beside the frames while the tank, bunkers and other components lay in a pile nearby.

The "Wren" class was the smallest in size yet numerically the largest of the Kerr Stuart's standard range of locomotives, and a total of 167 were built between 1905 and 1941 (the last four were constructed by Hunslet's who had obtained Kerr Stuart's goodwill). Two outside cylinders 6" by 9" drove the rear coupled wheels, 1'8" diameter. The boiler contained 35 1¾" tubes and generated steam at 140 p.s.i. Heating surface was: firebox 13.9 sq.ft., tubes 72.1sq.ft., total 86.0 sq.ft., and grate area 2.19 sq.ft. The water capacity was 87 gallons. In working order the locomotive weighed 4 tons 3 cwt. and had a tractive effort at 75% BP of 1700 lb. Overall measurements were approximately 9'6" long, 4'3" wide, 7'8" high and wheel base 3'. The first "Wrens" had inside Stephenson link motion, but most "Wrens" built after 1915 including "Haig" and "Pixie" were fitted with outside Hackworth valve gears.

A Motor Rail diesel locomotive was obtained and handled the majority of the traffic. One informant alleged it was new, but M.R. records do not show any sales to this company of either locomotives or spares, so presumably it was secondhand, but in very good condition. After the closure, it was sold in 1949. One source said that this locomotive was hired from Diesel LocoHirers which would explain the absence of records and its early disappearance after closure.

QUARRY SYSTEMS NOT CONNECTED TO L.B.L.R.

Two secondhand cabless Ruston & Hornsby diesel locomotives were obtained from John Heaver Ltd., Chichester. These were Nos. 172902, 16HP, built December 1935 and 174545, 18/21 HP, built May, 1937, both delivered new to John Heaver Ltd. They were sold to Frank Isaacson Ltd., West Drayton in 1952 and were observed in their yard in 1953. At Potsgrove only one of them had done any work. There was no locomotive shed - engines not in use were just sheeted over.

Rolling stock comprised about a dozen Hudson side tipping wagons, at least one of which had a screw brake. There was also an iron end tipping wagon with a scoop shaped body and inside frames.

An interesting relic which lasted until about 1962 was the body of a West Coast Joint Stock 6-compartment corridor coach with coupe at one end. At each end was a lavatory, and the opaque glass of the windows had the coats of arms of the owning companies - W.C.J.S. in one and L.N.W.R. in the other. The coach was used as a store and shack for the quarrymen. Bodies such as this were sold (less bogies) by Wolverton Carriage Works at £1 . per foot length.

QUARRY SYSTEMS NOT CONNECTED TO L.B.L.R.

Harry Sear, Manor Farm, Eggington

For a few years in the 1930's, Harry Sear, farmer of Manor Farm, Eggington, worked a small sand pit on his land (959255). It opened in 1931 or maybe 1930.

The pit produced loamy sand for building. It was dug by hand and loaded into 2' gauge wagons, which were pushed to a staging by the gate opposite the spinney above Eggington House, where the sand was tipped into lorries. Much was collected by Luton firms. The staging remained up to about 1940. A horse may have been used to haul the iron side tipping wagons (1 cubic yard capacity) of which there were about six. They were probably built by Hudson. The sand was screened during loading by throwing it through a portable framework mounted on the wagon.

In 1969 about 30 yards of track were exposed along the western lip of the quarry. The rail was $2\frac{1}{8}$" high and the gauge, measured at several sleepers, was $24\frac{1}{4}$" - presumably 24" with worn rails.

QUARRY SYSTEMS NOT CONNECTED TO L.B.L.R.

Woburn Abbey Parkland Railway

During 1968 a 60cm (1'11⅜") gauge railway owned by Mr.T.Barber was operated in Woburn Park (Grid. ref. 966328). About quarter of a mile of single track ran northwards from a station beside the main car park close to the House, more or less parallel to a park road.

The station had a gravel ground level platform enclosed by white palings, with Woburn Abbey on a raised board half way along. There was an entrance at each end. Opposite the platform was a steel water tank on a simple steel tower, and a pair of short sidings, the straight one being used to store the second train. Coal was kept beside the water tower. The light rail was bolted to steel sleepers laid on earth or sand ballast across the park. There were no earthworks, track being laid after removing the turf. No loops were laid; the locomotive pushed its train outward and pulled it back. No buffer-stops were fitted at the end of track.

Two locomotives were provided, one steam for use mainly at weekends and a diesel for other times. No.1 was an O-4-OST built in 1922 by the Hunslet Engine Co.Ltd., Leeds (wks.no.1429) for the Dinorwic Slate Quarries Co.Ltd., Llanberis, who had an extensive 1ft. 10¾ins. gauge system. Hunslet supplied many small O-4-OST to the Welsh slate quarries from 1870 onwards and No. 1, with two sisters, was the last of the line (the last was 1709 in 1932). The most noticable differences from the earlier locomotives were the provision of a cab, a domed boiler and deep buffer-beams since the frames were not tapered at the ends.

The boiler, with a heating surface of 100 sq.ft. (tubes 86 sq.ft., firebox 14 sq.ft.) supplied steam at 160 lb.sq.in. to two outside cylinders 7in. diam. x 10in. stroke which drove the rear axle. The wheels 1ft.8ins. diameter were inside the frames to permit a wider fire-box (grate area 2.5 sq.ft.) so fly cranks were fitted

QUARRY SYSTEMS NOT CONNECTED TO L.B.L.R.

to the ends of the axles to carry the coupling and connecting rods. The locomotive was 13 ft. long over buffers, 5 ft. 4 in. wide, 8 ft. 3 ins. high to top of the cab roof (7 ft. 3 ins. to chimney). In working order, with 100 gallons of water and 1½ cwt. of coal the engine weighed 6 tons 14 cwt., and developed a tractive effort of 2940 lb. at 75% boiler pressure. A set of drawings and photographs of this locomotive and its sisters were published in the March 1967 issue of "The Model Railway News". In the 1960's the Dinorwic rail system was progressively reduced and many of its locomotives were sold to enthusiasts. No. 1 was sold in 1967 for £1,000. It was overhauled by Gower, Bedford, who sent its wheels to Hunslet for regauging to 60cm. At Woburn No. 1 retained its Dinorwic maroon livery and cast number plates.

Rolling stock comprised two bogie open coaches and two 1 cu.yd. side tipping skips. The coaches had four doors either side and transverse slatted wooden seats. The frame was steel, with low truss rods, mounted on bogies from skip frames. Livery was pale blue with two horizontal black lines near the top, and a circular crest on the central side panel. The crest depicted a deer's head with a front view of a locomotive between its antlers. This device was surrounded by the words Woburn Abbey Parkland Railway.

Fares were 2/- adult, 1/- child, collected by a conductor after the passengers had boarded. Since the line was not fenced at all, the northern ends of the coaches, and the southern (bonnet) end of the diesel locomotive carried cow catchers. Side safety chains were also used to couple the locomotive to its coach.

The railway opened at Easter but at the end of the season it was dismantled and reopened in 1969 in Overstone Park, near Northampton.

W.D.L.R. 40HP SIMPLEX LOCOMOTIVE

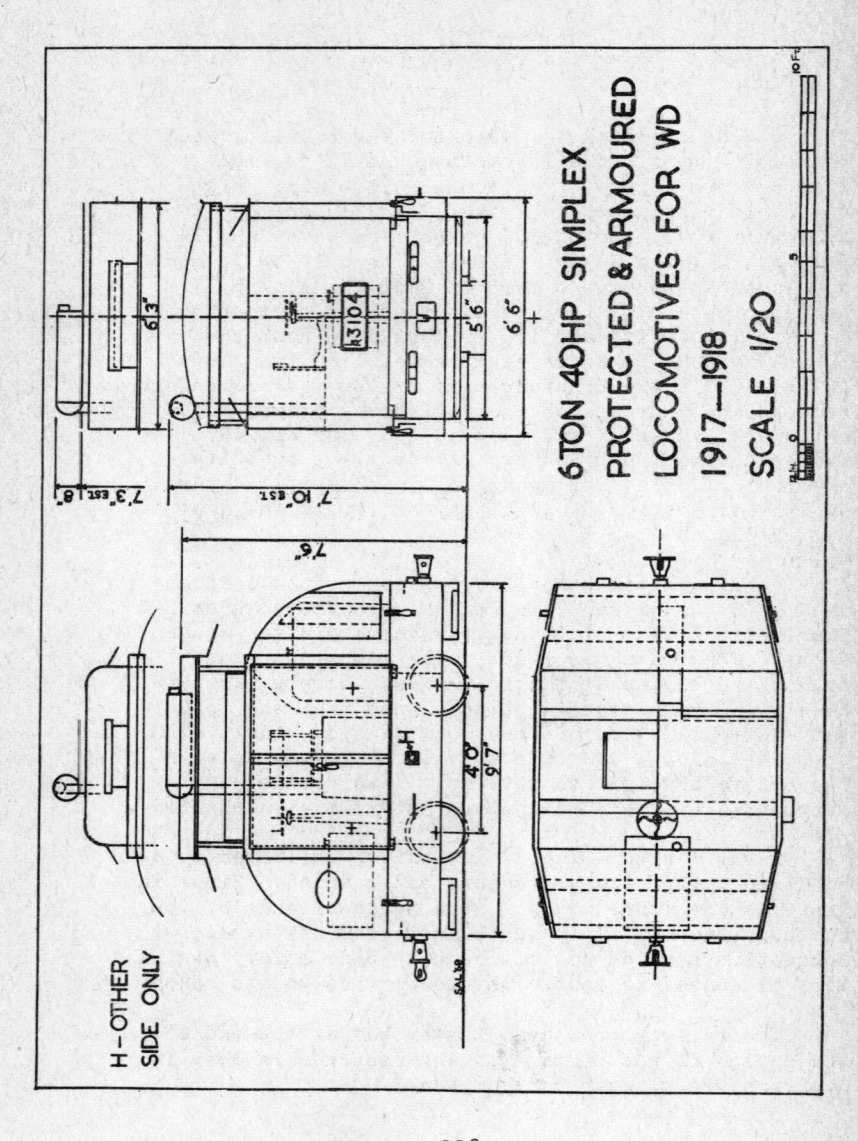

MODERN 40HP SIMPLEX LOCOMOTIVE

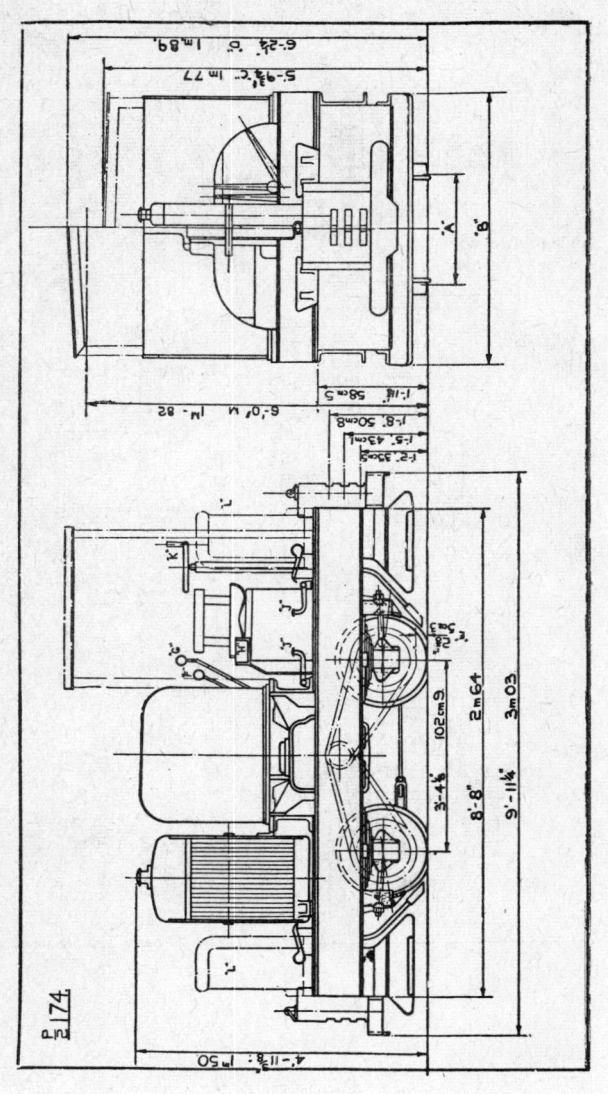

W.D.L.R. 20HP SIMPLEX LOCOMOTIVE

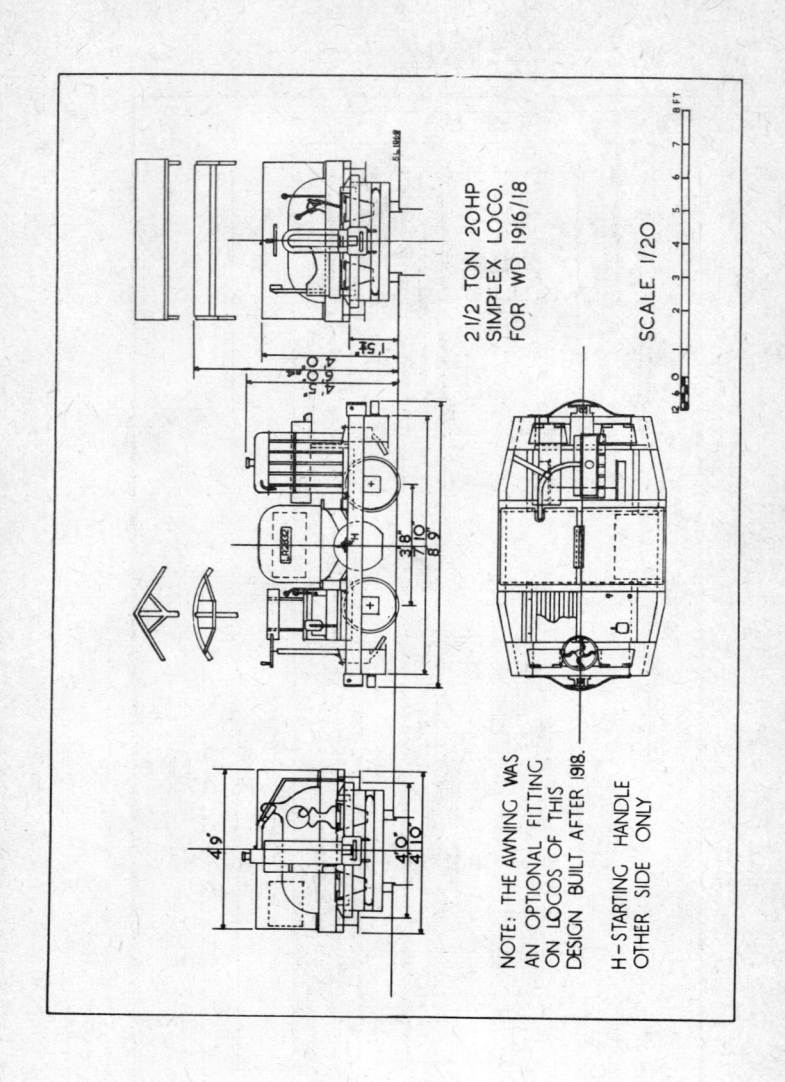

2 1/2 TON 20HP SIMPLEX LOCO. FOR WD 1916/18

SCALE 1/20

NOTE: THE AWNING WAS AN OPTIONAL FITTING ON LOCOS OF THIS DESIGN BUILT AFTER 1918.

H—STARTING HANDLE OTHER SIDE ONLY

MODERN 20HP SIMPLEX LOCOMOTIVE

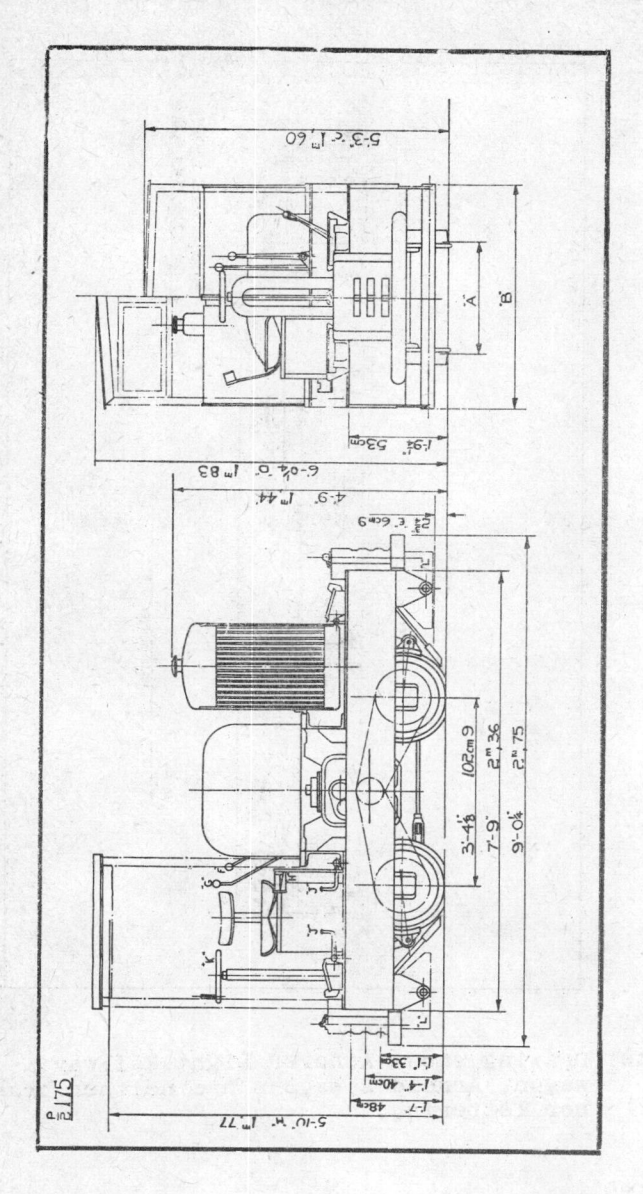

W.D.L.R. TYPE D BOGIE WAGON

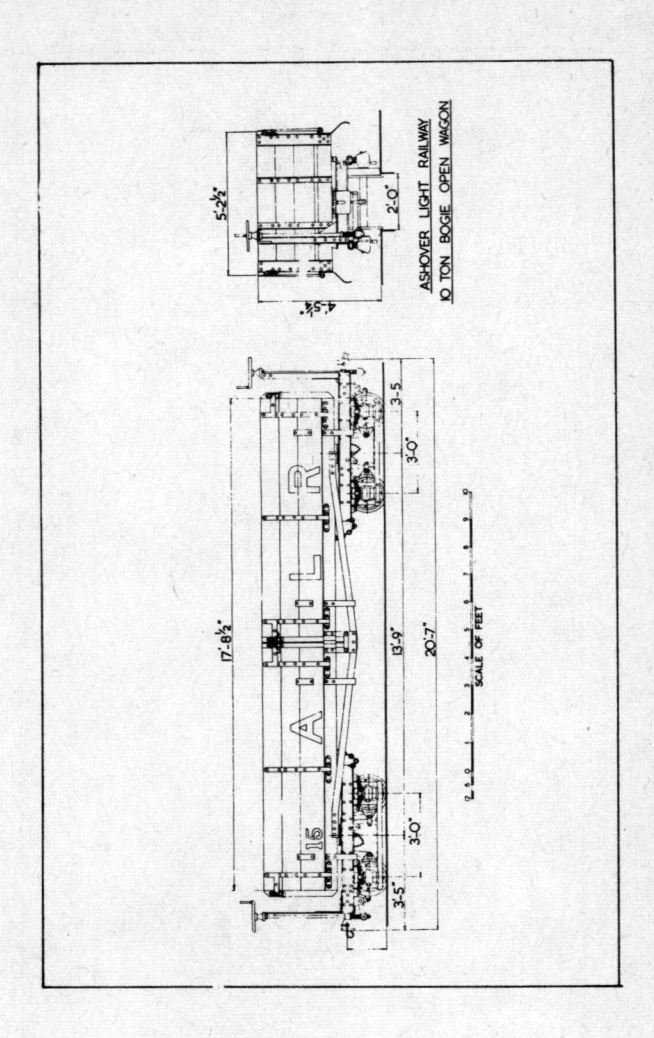

Note: Drawing shows Ashover Light Railway wagon. Arnold's wagons had neither brakes nor lettering.

HUDSON SIDE TIPPING WAGON

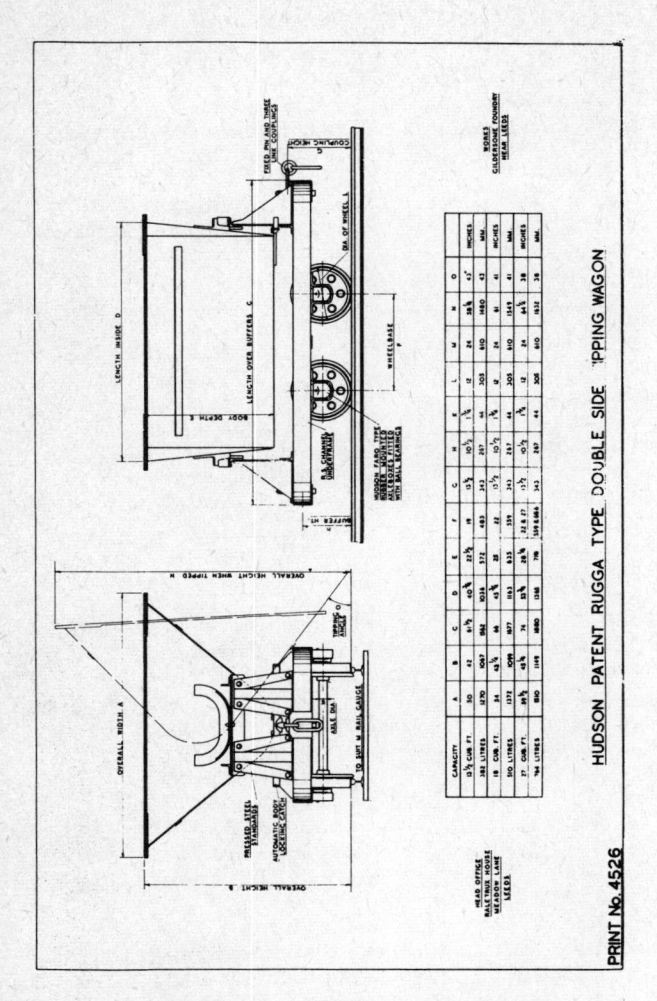

Note: End tipping wagons are similar but height increased to 53⅜", and 7" lengths of 4"x3" channel space the two underframes.

INDEX

Entries in **bold type** refer to drawings or maps.

Accidents 37-39
Air Compressor 53,69,91
Arnold,J.& Sons,Ltd.11, 14,17,19,20,22;24,27, 38,94,104
Auctions 7,20
Austro-Daimler 108
Bedford Silica Sand Mines Ltd. **4**,98-100
Billington Road Depot 11,12,13,24,**26**,27,35, 38,49-51,64-68,89,92, 94,101,104
Bridges 15,28,29,30,**31**,41
British Railways 22,24
Brown,H.G. 31,101
Chamberlain's Barn Quarry **4**,11,30,51,**52**,53,69,91
Clay Cross Coal Depot**4**,103
Construction 18,31
Contractor 18,41
Coaches 19,96,114,117
Crane 91
Delafield 12,17,70
Dividends 22,23
Double Arches Quarries
  Arnold **4**,11,24,25,31,47, **63**,64
  Garside **4**,13,25,31,**58**,60-62,69,88,110
Driroof Tiles Ltd.30,**52**,57
Eastern Way 13,24,31,**58**,60
Eastwoods Ltd.**103**,107-110
Eggington 15,115
Festiniog Railway 93
Firbank Pit **6**,104
Garside,G.(Sand) Ltd. 12,13,14,17,20,22,24,28, 104,107
Grand Union Canal 13,17,105
Great Northern Railway 15
Grovebury Quarries **4**,10,11 12,88,91,**103**,104-110
Harris,G.11,50

Heath & Reach 3,98
Horses 9,10,16,20,49,65,93 99,108,111
Iron Horse Preservation Society 25,28,51,89,93,94-97,113
Leighton Buzzard 3
Leighton Buzzard Brick Co.
  Ledburn Road Pit **4**,111
  Potsgrove Pit **4**,112
Leighton Buzzard Concrete Co.26,29,34
Leighton Buzzard Tiles Ltd. **4**,**103**,105,107,108
Leighton Buzzard Sand Co.104
Level Crossings 17,27, 29, 31,39,54,101,105,108
Locomotives
  L.B.L.R. 19,23,32,40-48,66 69,**118**,**119**
  Quarry51,54,57,60,65-86,99 100,107,108,109,112,113, 114,**120**,**121**
  Other 95-97,116-117
Locomotive Sheds & Workshops 22,23,27,28,31,43,44,50,51 53,60,61,62,64,94,106,107
Locomotive Stock Lists
  Arnold 72-79
  Garside 80-86
  L.B.L.R.36,47-48
London & NorthWestern Rly. 12,14,19,114
Marley Tiles Ltd.**26**,29,34, 38,39
Mars Bars 47
Midland Railway 14,15
Motor Rail Ltd. 41-43,45-47 65-67,69,**118**-**121**
Munday's Hill Quarry **4**,13, 25, 30,**58**,59,62
New Trees Quarry **4**,11,25,30 **52**,54,64
Nine Acres Quarry **4**,11,30, 55-57

Opening of L.B.L.R. 19
Operation 22,28,29,32,33-7,49
  50,54,55,57,62,99,110
Page's Park 4,28,35,50,04
Parrot & Jones Quarry 4,52,55,
  57,59
Paul,H 102
Pratts Pit 4,26,27,49
Quarrymen's Methods 8-10,111
Quarry Railways (General) 8,10,
  20,32
Rackley Hill Quarry 11,12,103
  104-105
Railway Schemes 6,7,14-20
Readymix Transite 106,110
Redland Tiles 17,108
Roads 6,7

Sand 3,6
Sand Carting 6-8,20
Sear,H 4,115
Signal 37
Snowploughs 39,92,110,126
Spinney Pool 11,104
Stonehenge Brickworks 4,25,30,
  39,52,55-57
Track 16,19,20,22,29,31-32,41
  54,99
Tractors 7
Twenty One Acres Quarry 4,55-6
Wagons 20,33,56,87-92,99,102
  109,110,114,117,122,123,126
Wagon Allocations 93
Wayleaves 18,20,22,25
Winfield,H 102
Woburn Abbey 116-117,124

Late Information

p.54: Chamberlain's Barn - Main Line connection lifted
near washer c.6/69 due to expansion of Anchor Tile
Works. Quarry now isolated from L.B.L.R.

p.79: sale of Arnolds locomotives: 2 to Lithgow, Glasgow
2/9/69;3 to T.Gray & Sons Ltd. Burton Latimer 13/8/69.
Woburn Abbey: No.2 was a Motor Rail diesel,No.8993,
built in 1946 for Sir Robert McAlpine and obtained
from Willett Engineering, Snodland sewage contract.
It was black, with a new cab and rectangular bonnet
made by Mr. Barber

ADDITIONAL INFORMATION

Iron Horse

Regular passenger trains are now run on Sunday afternoons. Visitors receive a "pass" which entitles them to one return journey on the railway "having given a donation". This avoids the necessity for an M.o.T. inspection. The usual locomotive is "Pixie" resplendent in green livery, lined in black and yellow. A raised wood platform 46' long has been built at Page's Park beside the southern track and a water tank mounted on a tower of sleepers nearby. The tank is filled with water brought by train in tanks filled at the tap at Marley Tiles as a mains supply is not yet (Aug.1969) laid on. Several more wagons have been acquired including two bogie side tippers, one of which has already been converted to a coach. The crane (103) may be mounted on the chassis of loco. 5613 to improve its stability. One of the other engineless Simplex locos. is used as a brake van at the northern end of passenger trains.

Wagons

Both Arnold's and Garside's used to have their own standard gauge wagons. Arnolds had thirty in red oxide livery and Garside's forty-eight, half in amber and the rest in black livery. Arnold's were not accepted by B.R. in 1947, while Garside's had been absorbed into the common pool in 1939. Between 1950 and 1957 B.R. built 1000 steel 13 ton wagons for the Leighton sand traffic.

Arnold's snow plough is stored at Billington Road. It is blunter and deeper than Garside's.

Private Railway, Heath

A house at the end of Eastern Way (926279) has about 60 yards of 2' gauge track and a side tipping wagon to carry wood from a yard to the house's boiler.

Double Arches (Arnold)

North Pit reopened Dec.1966, with a face very close to Garside's Quarry.

CORRECTION

Arnold's locomotive 9409 (f/n rr): delete "Scr./66", substitute "to Stone Pits Aylesbury 29/11/66 as No.8 to replace an earlier No.8 sent from Billington Road 8/7/59.